Advance Praise for *Housing Reclaimed*

Jessica Kellner's book comes to us in the nick of time. We need a new, more enlightened approach to housing and this book provides the roadmap. *Housing Reclaimed* could put a whole generation on the path to comfortable, secure sustainability. Jessica has written a beautiful and necessary book that everyone who lives under a roof should read.

— Bryan Welch, Publisher, *Mother Earth News*,
Natural Home & Garden and the *Utne Reader*,
Author, *Beautiful & Abundant: Creating the World We Want*

In an environment of underwater mortgages, home foreclosures, and lack of adequate housing for many Americans, Jessica Kellner's *Housing Reclaimed* makes a compelling case that we can more easily realize the dream of home-ownership if we utilize our hands, our imaginations, and the high-quality low-cost materials available from building deconstruction. Filled with many creative and innovative examples of warm, livable and affordable homes built from found materials, this book should be in the hands of anyone who wants to build his or her own home without getting trapped by the large debt associated with conventionally marketed and financed houses.

— Bob Falk, President, Building Materials Reuse Association,
Author, *Unbuilding: Salvaging the Architectural Treasures
of Unwanted Houses*

In a time when so much of the news around housing is negative, Jessica Kellner offers an optimistic but practical approach to building a home — mortgage free! Jessica proves that, with a little creativity and a willingness to step outside the constructs of modern housing, anyone can build a dream house.

— Robyn Griggs Lawrence, Author, *Simply Imperfect:
Revisiting the Wabi-Sabi House*

This unique book outlines an inspiring perspective on how we can make housing more sustainable and more affordable. Kellner provides compelling examples of how we can build our own elegant, debt-free homes, and she outlines approaches to make housing sustainable by creating salvage businesses, showcasing companies that recycle entire homes and non-profits that produce sustainable low-income housing.

— Cheryl Long, Editor-in-Chief, *Mother Earth News*

Jessica Kellner has managed to give us a glimpse of who we are as a species — clever, creative and resourceful. Perhaps we can take a hint and return to primal sensibilities and first strategies, and discover who we really are. She even tells us where to go to do that. Magnificent!

— Dan Phillips, Founder and owner, The Phoenix Commotion

In *Housing Reclaimed*, Jessica Kellner ventures into terrain that remains off-limits to most: the subculture of homes made from trash, reclaimed, discarded and recycled material. In exploring case studies of people who have crafted their homes out of society's cast-offs, Kellner challenges all of us to think outside the box of residential convention and to embrace new options. Anyone interested in saving a buck, in saving the planet and in creating a magical, healthful and one-of-a-kind home should reach for this beautifully crafted, engaging and timely book.

— Wanda Urbanska, Author, *The Heart of Simple Living:
7 Paths to a Better Life,* Co-author, *Less is More:
Embracing Simplicity for a Healthy Planet,
a Caring Economy and Lasting Happiness*

HOUSING **RECLAIMED**

HOUSING **RECLAIMED**

Sustainable Homes for **Next to Nothing**

Jessica Kellner

NEW SOCIETY PUBLISHERS

Cover design by Diane McIntosh.
Image © Veer

Printed in Canada. First printing 2011.

Paperback ISBN: 978-0-86571-696-4
eISBN: 978-1-55092-493-0

Inquiries regarding requests to reprint all or part of *Housing Reclaimed*
should be addressed to New Society Publishers at the address below.

To order directly from the publishers, please call
toll-free (North America) 1-800-567-6772,
or order online at newsociety.com

Any other inquiries can be directed by mail to:

New Society Publishers
P.O. Box 189, Gabriola Island, BC V0R 1X0, Canada
(250) 247-9737

New Society Publishers' mission is to publish books that contribute in fundamental ways to building
an ecologically sustainable and just society, and to do so with the least possible impact on the
environment, in a manner that models this vision. We are committed to doing this not just through
education, but through action. The interior pages of our bound books are printed on Forest Stewardship
Council-registered acid-free paper that is 100% post-consumer recycled (100% old growth forest-
free), processed chlorine free, and printed with vegetable-based, low-VOC inks, with covers produced
using FSC-registered stock. New Society also works to reduce its carbon footprint, and purchases
carbon offsets based on an annual audit to ensure a carbon neutral footprint. For further information,
or to browse our full list of books and purchase securely, visit our website at: newsociety.com

LIBRARY AND ARCHIVES CANADA CATALOGUING IN PUBLICATION

Kellner, Jessica
Housing reclaimed : sustainable homes for next to nothing / Jessica Kellner.

Includes index.
ISBN 978-0-86571-696-4

1. Housing — Finance.   2. House construction.   3. Building materials — Recycling.
4. Ecological houses.   I. Title.

HD7287.55.K45 2011   333.33'82   C2011-904419-6

NEW SOCIETY PUBLISHERS
www.newsociety.com

MIX
Paper from
responsible sources
FSC® C016245

Contents

Acknowledgments

Thank you to the many people who contributed in various invaluable ways to the writing of this book, in particular: James Duft; Mike, Laura and Beth Kellner; Bryan Welch; Robyn Griggs Lawrence; K.C. Compton; Fred Robertson; and Cynthia Dodd.

Introduction

Our homes are the most intimate of spaces; the backdrops of our lives. The need and desire to create a shelter for family and self is as ancient as human civilization itself.

For most of human history, we have created our homes with our hands, out of the materials available to us where we live. We've altered our homes as our families have changed. We've designed them for ourselves and our lives. We've formed communities around them.

Since the Industrial Revolution, our homes have become increasingly alienated from us, and we have alienated ourselves from them. As our professions have become more specialized and our lives more compartmentalized, mass production, increased access to credit and layers of bureaucracy have carried us farther and farther from the path of self-sufficiency. Today, our food is shipped from thousands of miles away, and our homes, especially our low-income ones, are quickly constructed, uniform boxes designed for everyone, not anyone in particular, using often-toxic, low-quality materials.

At the same time, the invention and standardization of the 30-year mortgage and our ever-increasing reliance on the credit system has come to mean that most of us never own our homes outright. In many cases, all we pay is interest to the bank, confident that ever-rising home values will eventually lead to a financial gain in the risky housing market. Rather than investments in one's family and future, houses have become financial investments, valuable not as a place but as a commodity.

The need for home prices to climb continuously has edged out many low-income families, who simply can't afford even the lowest-cost homes on the market. The need for home prices to climb continuously was also the underpinning of the subprime mortgage crisis of 2006 to 2008 — and proof that participation in the conventional home market is riskier than most homeowners believed. The value of housing, having become a stock market commodity, was allowed and encouraged by the free market to increase in value far beyond its worth in wood, concrete and nails. Irresponsible lending, greed, ignorance and government deregulation worked together to ruin the financial lives of millions of Americans and to cause millions of others to lose their homes.

And as home quality has gone down and home prices have gone up, our throwaway culture and the throwaway housing market have increased our waste to astronomical levels. We demolish more than 250,000 homes a year, according to the National Renewable Energy Laboratory.[1] Almost all of the building supplies in those homes are bulldozed, crushing everything inside and sending it to the landfill. We send more than 135 million tons of construction site debris to landfills every year, according to the Environmental Protection Agency.[2] Of that, half is demolition waste, and 40 percent is from renovation and remodeling. Meanwhile, every urban area in the nation, as well as many suburban and rural areas, has shortages of affordable housing. In 2010, the National Low Income Housing Coalition released a study[3] that showed that, from 2007 to 2008, growing demand and shrinking supply of affordable and available rental units for extremely low-income households led to an increase in the absolute shortage from 2.7 million to 3.1 million homes.

It's time we begin thinking differently about housing, in terms of what our shelters are and should be made of, and of how we create and inhabit them. Housing isn't meant to be a one-size-fits-all, bigger-is-better proposition. Today, all over America and the world, individuals and groups are creating homes that don't fit the mold. Homeowners in Alabama, Idaho and Colorado are creating small, artful homes using salvaged materials, never taking out construction loans. In Texas and North Carolina, people are working together to reclaim building supplies and whole houses before they go to the landfill, using them to create new homes and neighborhoods for hardworking families. In Reno, a pair of designers, sick of seeing their inner city crumble, is revitalizing old buildings and blighted neighborhoods.

The individuals and homes featured in this book show the ways that regular people have extricated their homes and their communities from the standard model. Rather than designing their homes for the real estate market, these brave individuals designed their homes for their own lifestyle. They reject the "bigger is better" mantra. They reject "resell value." Their methods of simplicity, reuse and community-building engage our deepest connections and relationships with our homes. In place of mortgages, they invested time and love. Instead of connecting over a shared desire for three bedrooms and two-and-a-half baths, they connected over their shared desire for community. Rather than hosting housewarming parties, they hosted mudding parties.

In this book, I hope to show that building homes out of reclaimed materials is an idea that applies to a whole lot of people. You can do this if you are the type of person who wants to get involved and build your own home. Building your own home isn't crazy. It's something nearly every person used to do not so long ago. It's not impossible today, and it doesn't have to be intimidating, even in the city. The skills required to build a home are, for the most part, simple and easy to learn. It makes good financial sense to create homes that cost less, too. It makes even more sense to use all the quality building materials we currently send to the landfill simply because we haven't figured out a better destination for them. Programs featured in this book model the ways we can access used building supplies and unconventional building methods to provide low-cost housing and improve communities.

We face challenges today that have never existed in the past. Our growing world population means the need to provide low-cost housing will expand each year. Our serious environmental crisis means we must be vigilant about creating homes that run efficiently. But we also have many advantages. Technological advances, design evolution and resourceful, out-of-the-box thinking in terms of materials and efficiency can help prepare us to meet challenges never before faced.

With challenges comes the opportunity for great growth. Our history is one of great, almost unthinkable advances in times of need. Great revolutions in thinking and practice enable us to create new realities when it seems we've hit a dead end. Our recent economic crisis can light the fire to change the commoditization and depersonalization of our homes. It's clear to see our system needs change. By changing the way we conceive of and demand our homes, we could start to change the system.

We are not slaves to the machine of mass-market housing. The skills that enabled our great-grandparents to build their own homes are still buried within us. The desire to create safe, beautiful homes for ourselves and others is strong. And it is our human nature to work together and help one another better our lives. We have plenty of materials and many examples. In this book, I hope to inspire you with views of amazing homes, lists of resources, stories of intense human spirit and practical examples that prove we can take back our right to housing. We can make our homes places we value because of the lives lived within them, not places we value because of the mortgage that hangs over our heads. We can connect with our homes more deeply because we know how they were built and where every building material came from. We can provide more for ourselves and our families because we own our homes, mortgage-free. We can use all the quality building materials available to us to build homes for our neighbors and our communities. And we can also use all those valuable supplies to reinvigorate urban centers and provide the very low-cost housing we need in increasing numbers. We can reclaim our right to housing.

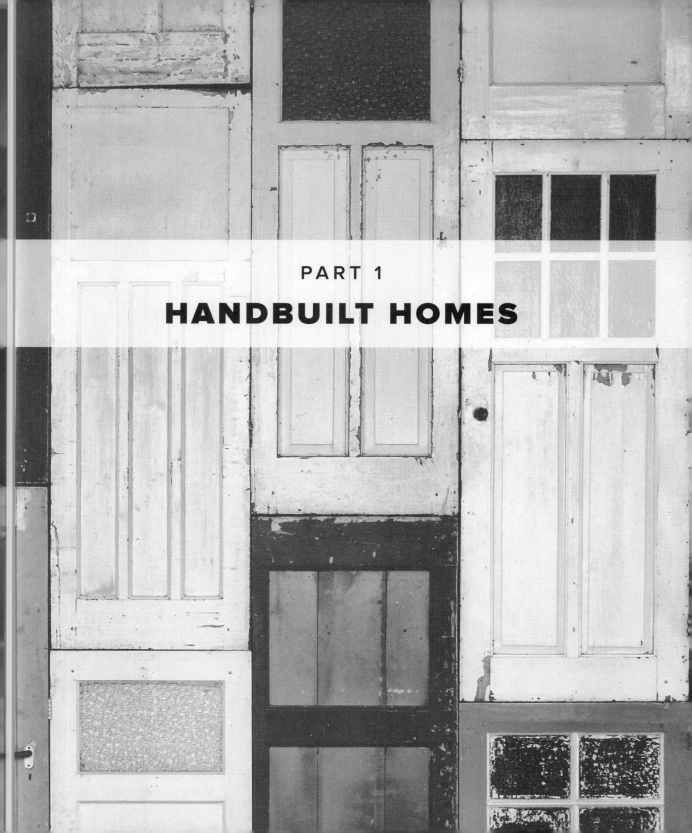

PART 1

HANDBUILT HOMES

CHAPTER 1

ALL IN THE FAMILY

An Alabama family comes together
to hand-build a home and connect with
each other and their region's history
in the process.

IN WEDOWEE, ALABAMA, Guy and Kay Baker live in a cozy cottage they built with their three sons using almost entirely salvaged materials collected from all over their county. Under the guidance of Guy, a lifelong professional builder, the family spent about five years on the project, lovingly and painstakingly building the intimate space using centuries-old materials. The family so loves their handbuilt home, initially planned as a vacation cottage, that they ended up moving in full-time, and every day Kay and Guy enjoy the personal connection they have with every detail of the 1,100-square-foot space.

In 2001, Guy was overwhelmed at work, and Kay was working on her bachelor's degree in psychology. The couple's three young sons were getting increasingly busy with school and personal lives. When Guy's mother unexpectedly fell ill and passed away, Guy became acutely aware of the sensation that life was passing him by. He felt driven to make good on a longtime dream of building a getaway in the woods for himself and his family.

Kay and Guy had owned the land on which they planned to build — formerly owned by Guy's grandfather — for years, but they'd never gotten around to starting the project. Eager to reconnect with his past and the things he values most in life, Guy was inspired to get moving on the project after his mother's death: "It was something I had always wanted to do, but I'd always put it off. Things just got in the way—work, school, the boys. We were just making excuses for never doing it. But my mother had gotten sick in 2001 and passed away, and I think that was the reason I went ahead and quit making excuses and just found the time to do it."

For Guy, his family was building more than a home; they were building a place for calm and family togetherness, a place to escape the hectic world. "The biggest reason for doing it was that my workload had gotten astronomical. I had no down time, and with the boys at the age they were, we just needed some peace and serenity," he says.

The Bakers' 1,100-square-foot cabin is made with 85 percent reclaimed materials the family collected from all over Randolph County, Alabama.

■ Building Reclaimed

Guy had long had a fascination with the array of antique building materials he saw while working on tear-down buildings in the area. He was impressed with the materials' good quality and durability, even after they had withstood the elements for hundreds of years. He saw the antiques he'd collected as heirlooms of a bygone era that valued craftsmanship over speed. "I was always and still am fascinated with older structures and older materials. It amazed me that I could work on houses that were 150 years old, and the damages to these homes were minute because of the materials and the quality of the studs and the lumber," he says. "A year later, you work on a home that's only 20 years old, and you saw all this termite and water damage."

Guy and Kay estimate they spent $20 on the kitchen; hinges and doorknobs were the only things they paid for. Kay laid the wood-block countertops herself.

For years, Guy had been collecting items — bits of the region's architectural history — gathered from projects in the area. Though he hadn't been sure at the time what he would do with them, he knew those great old things were too wonderful to throw away. When it came time to start construction on his family cabin, Guy realized he had probably collected nearly enough reclaimed materials to build the whole cabin. He knew the unique materials would give his home a one-of-a-kind feel. "You couldn't purchase the boards in this home nowadays," Guy says. "Even if you tried to duplicate it, you couldn't. They're all one-of-a-kind. Most of the boards were hand-hewn with a chop-axe, and they were in excellent condition. That was fascinating to me: to be able to take something that had been out in the elements for hundreds of years and it was still good quality."

An old Indian grindstone the family found on the property serves as decor in the outdoor kitchen supply building.

Having spent 20 years building in Randolph County, Guy had more than his collection of antiques to call upon when he started building his own home. He also had a vast knowledge of all the area's best sources of reclaimed and antique building materials. "All these materials were readily available. They were everywhere," he says. People in the area who were tearing down old structures often didn't have another destination for them, so Guy took them off their hands. "Being in the construction industry, I saw it everywhere — say we were tearing down an old barn with great old wood. If you ask them if you can have it, nine out of ten people say, 'Sure!'"

Guy's collection of building supplies and his knowledge of how many additional resources were available in the area helped convince Guy and Kay to build the home in the first place. The free materials made building their home a low-cost endeavor. "After being in this business for years, I knew I could build what I wanted at a very minimal cost, and we did," Guy says. He and Kay were determined to avoid taking on debt to build their dream home. Over time, finding free supplies became a game to Guy. "It came to the point that you didn't want to spend anything," he says. "Anytime you needed something, you knew it was out there, and you could find it. It almost became a challenge to not spend any money and be able to do this."

Guy searched far and wide to find the best materials to use in his home, then used creativity, artistry and hard work to incorporate them into his home. He used entirely antique window panes from an 1800s church his company worked to deconstruct. People said the church was the oldest in the county, and Guy spent hundreds of hours reframing the antique panes with reclaimed wood. He estimates creating new window frames from reclaimed wood and fitting the panes took him about 60 hours per window — and there are 12 windows in the house. He created a gigantic bathtub by lining a cattle trough with fiberglass. Guy estimates he spent $20 on the kitchen — doorknobs and hinges were the only thing he paid for. The outdoor stone fireplace is made of stones collected on the property. Old road signs and American Indian grindstones found on the property act as decor.

Though safety required that some building materials such as plumbing and wiring be new, overall the project cost virtually nothing. "Other than the wiring and the plumbing and things like that, we didn't spend any money," Guy says. He was able to hunt down just about everything they needed from the many old buildings in their rural area. And though saving money was part of the motivation, the family was also keen on using reclaimed materials because they liked incorporating their region's history into their dream home. "In this part of the country, farming was the main industry, so barns are everywhere. They're dilapidated, but there's a lot of good lumber in those

Building the home together brought the Baker family closer and instilled a huge sense of confidence and can-do spirit in the Baker sons. Left to right: Jeffery, Adam, Kay, Guy and Kyle.

things," Guy says. "It was partially monetary and part just being fascinated with the idea that you could take something someone built 100 years ago, take it apart and create your own dream."

■ The Ultimate Family Project

As Guy collected materials (he says he used something from every town in Randolph County), the Baker family started spending their evenings and weekends building. From roofing and tiling to laying flooring, the family members took on every task. The boys, Jeffery, Kyle and Adam, who were 15, 14 and 12 when the project started, were assigned specific jobs, such as constructing the outdoor fireplace from rocks found all over the property. Youngest son Adam says he touched every one of the thousands of rocks used in the outdoor fireplace, foundation and dry creek beds three times — once when he found a stone, once when he moved it to the house and once when he laid it in its final destination. Kay tiled the kitchen floor and countertops with "tiles" made of barn wood Guy cut into small slices. She then coated each one with polyurethane for a shiny effect. "The wood was from barns and floors," Kay says. "Guy took it and sliced it like a loaf of bread. They were different sizes. Then he told me how to pattern them and lay them out as tiles. We did that on the countertop and the floor. It's a little crooked, but it's still beautiful," she says.

Guy and Kay viewed building their home as an important way to teach their sons the value of hard work and to show them what a huge feat they could accomplish working together as a family. Working hands-on together provided the family a way to connect outside the normal day-to-day grind, and it taught Kay and Guy's sons the importance of dedication.

Though the teens may have grumbled at times as they made their way through the project, today all three Baker sons realize the invaluable lessons they learned from building the home, and they know their home is worth all the hard work. Jeffery and Kyle also gained a foundation for their careers through the project — both are professional builders working for Guy today. Middle son Kyle says that, though the project was challenging, he gained enthusiasm as the home came together: "At first, I hated it. That's the last thing you want to do with all your free time when you're 16 and 17 years old. But later on, all the pieces of the puzzle came together. When it got closer to the finished product, you saw how neat it was and you wanted to do more to it." He remembers how the project empowered him, and it's still a source of pride: "I remember rocking the inside of the fireplace. I pretty much did that by myself. I'd never done anything by myself. I'd probably never been trusted to, but I did that. I'm proud of it." Kyle says the project helped him determine that

he wanted to be a professional builder, but also that he could do anything he set his mind to: "It showed me that I was able and capable of doing some of this stuff, and now I'm getting paid for it. It opens your mind to all the things you can actually do if you just get down and try it."

Collecting stones from all over the property and laying the outdoor fireplace with them were some of the Baker sons' tasks.

Adam says working with his parents helped him develop a stronger, more mature relationship with them: "There is a certain amount of time you have to spend with a person before you truly know them. Until you've seen your parents react to frustration in a very human way, you can't say for sure what type of person they are." Watching as their parents demonstrated grace under pressure showed the Baker sons the best way to react to difficult challenges, Adam says: "I had the opportunity to see how both of my parents deal with stress, and it built respect between us. They did not have to tell me to be moral or honest; they showed me by example. They did not throw a fit every time a rock refused to stay cemented to the wall. We saw what was happening and learned from it."

Along with the value of accomplishing goals, Adam admits the project also taught him the value of failure, and of perseverance: "If I succeed without making any mistakes, I fail to learn something new. What is an accomplishment if we've learned nothing from it?" And, though he learned a lot about his parents and brothers, working through the challenging project also helped Adam learn about himself: "Everyone has character flaws. My family loves me enough to point mine out. They've helped me become a better person by showing me things about myself I couldn't see on my own." Adam says he grew up while building the project from ages 12 to 17 and that it helped him navigate the sometimes difficult road to adulthood: "Every one of us changed while building that place, but we grew together, not apart. It was therapy to get our minds back on track after the pitfalls of everyday life. Building that house helped me cope with the death of my grandmother and two close friends. I became a man. I grew closer to my father. I got to know my brothers. I learned to appreciate my mother. She is my link to the past and what keeps me moving forward." Adam also built the foundation of what he hopes is his future career — he's applied to architecture school at the University of North Carolina in Charlotte.

Kay agrees that everyone in her family grew through their work on the project. Building the home helped her sons become the responsible, capable adults they are today, she feels: "This project just instilled something inside of them. All three of the boys have such a huge sense of independence, and it seems to have come from this project — working like they did, realizing they could make anything happen. It made us really close." Guy sees the home as a testament to his family's dedication: "I love that when my sons come in this house, it's a reminder of what hard work and dedication can do. At the time, they weren't crazy about doing it like I was. But now the boys can remember doing each task, and it becomes really personal. It's more than a house, when you know that you or your children had your hands in the whole project."

Kay loves how every part of the home has its own family memories — not just the memories of the times they've spent in it over the years, but of actually laying the bricks together. The entire family recalls fondly constructing the large fireplace that extends up through the family room and heats the entire home, a task that was hilariously wrought with difficulty. Kay remembers: "Oh, the fireplace, I just laugh when I look at it and think about it. We built that on one of those days when it was freezing cold, and probably 10 degrees colder inside than outside. All the boys were here, and Guy was trying to get all the bricks up there. Our middle son was on the ladder, and he was mad because it was really difficult, and he was holding the mortar. Every time he'd try to put it up on the bricks, he'd sling some of it on my head. Of course, I'd get mad, but we all ended up laughing. It's a fun memory."

The project made the family so close that sometimes they didn't have to talk at all, Adam says: "I remember days where no one said a word because we had had an argument, and the thing that broke the silence was what we had built that day. We were so close and knew each other so well that we could build anything together and speaking was an option. You don't learn that much about someone just by having dinner together one night a week."

■ Home Sweet Home

When their sons were young children, the Bakers lived in a number of homes. Guy was continuously buying and renovating homes, and the family would often relocate into his least finished project while he fixed it up. Living in a number of places, Kay says her sons never developed a strong sense of connection to a place until this home: "Of the 15 homes my sons have lived in in this world, this is probably the only place my sons really think of as home." The home is more than a house to the Baker family. It's an heirloom and a repository of family memories. "The idea's been batted around about selling it," Kay admits. "We'll say, 'If someone pulled up in this driveway and offered however much money, would we sell it?' The boys look at us and just say, 'No.' This is the one place that means a lot to all of us. It's something they can hold onto and have."

Oldest son Jeffery confirms that, though they've lived in many houses, the little cabin will always be home to them: "I don't think it would matter where any of us lived, that would always be home. We would always want to go back

Though they've lived in many houses, the Baker family says this house is the only one that truly feels like home because they put so much love and effort into it.

Michael Shopenn

there for Christmases and all that." All three sons realize their connection is so strong because they put so much work into it. Adam concurs: "My family's house feels so much like home to me because I had a hand in it. The average homeowner is two degrees away from their home. They work a job to earn money that pays for labor and material cost. Often they will help with the design of the house, but they will never know the feeling I got from clearing nails from old boards knowing they would be given life again. I worked on my family's home with my own hands. Money isn't what's needed when you do something like this — the main thing you spend is time with people you love."

■ Guy's Four Requirements for a Handbuilt Home

Guy says having patience is the top requirement of taking on a project like this: "I learned at a young age that patience is the key to anything. Patience and

Guy built the bed frame out of reclaimed timbers, modeling it after an expensive one Kay saw in a magazine.

Michael Shopenn

knowledge: If you have those two things, you can pretty much do anything. The biggest key is realizing you're not going to do it overnight." Before starting the project, Guy accepted that it was going to take a long time, but that he would get through it by focusing on what he and his family had accomplished rather than what was still left to do. He says concentrating on achieving one small goal at a time helped keep the project manageable. Going slow also gave the family the opportunity to see their progress as they went along, and to really contemplate how they wanted to build every detail. Guy instilled the importance of patience in his sons, teaching them to enjoy the process as much as the finished product. Adam says: "I wanted to complete a task by doing everything at once. My father taught me to do things one at a time so I could visually see my accomplishments. Otherwise I would get frustrated and want to give up. Now, I have learned to enjoy using my hands and building something. It gives me time to think."

Guy made the bathtub by lining a cattle trough with fiberglass; the whole project cost $90.

Knowledge is the second requirement on Guy's list, but he feels that anyone willing to investigate and learn could figure out nearly every skill needed to build a home. Even as a builder with 20 years of experience, he says sometimes he had to learn a new skill along the way: "We all get to the point where there's something we don't quite understand, and when we get to that point, what do we do? We ask questions or we get online until we find the answer, then we move forward. It would be the same thing for someone who didn't have the knowledge or background. It would take them longer, but if that's what they want to do, they can do it."

Though Guy feels anyone could learn the skills to take on a project like this, that doesn't mean it didn't take a whole lot of his third requirement: hard work.

In the depths of the project, Kay recalls wondering if the project would ever end: "I was overwhelmed like it would never be over. It was such a long process. It took Guy every minute he had. He would work all day and come to the cabin for two or three hours in the afternoon." She laughs, "From a wife's point of view, I didn't think it would ever end."

The family's biggest investment into the project was the time they put into it. "It's time-consuming. Anytime you go with recycled materials, and you're doing it on weekends and after hours, it becomes very time-consuming," Guy admits. Though using reclaimed building materials helped reduce the financial

The Baker home is filled with salvaged wood, windows and decor from the region's architectural history Guy collected over the years. The windows throughout the home were salvaged from one of the oldest churches in the area; Guy painstakingly reframed each one by hand.

A Customized Home

One of the best parts of building your own home is the ability it gives you to incorporate your family's desires into its design. The Baker family loves being outdoors, and with Alabama's nearly nine months of warm weather every year, outdoor living is an important part of the family's life. They incorporated an outdoor kitchen with an outdoor storage supply shed that houses a sink and cooking tools. Guy explains that the outdoor fireplace, on a porch tucked into the surrounding woods, is one of his favorite parts of the home: "We wanted it to feel like the house was part of the living area, and incorporated it with the outdoor kitchen." Even in winter, Guy says he and Kay make good use of their outdoor spaces. He loves sitting next to the outdoor fireplace with a hot cup of coffee on one of the region's few snowy winter days.

The Bakers love spending time outdoors, and designed their home for comfortable, year-round outdoor living.

expense of the project, it increased the time investment of nearly every task. "It's not the simple task of calling the materials store and having stuff delivered," Guy says. "You're actually bringing the materials from somewhere else. With wood planks, you're denailing them and using them again. That becomes a task." But Guy feels it was important to him not to cut corners on the place he planned to live in for the rest of his life. He knew if he did, he'd always look at the rushed job and wonder "why didn't I just spend the extra 50 hours to make that right?" The family says all the added effort required to prepare the reclaimed materials was worth it for the one-of-a-kind home it created. "Whether you have experience or no experience, if you do something like this, it's one of a kind," Guy says. "Even I could not duplicate this cabin we live in because the materials were all unique. It's a one-of-a-kind thing."

■ Cabin and Community

Building the house with items collected from all over the region connects their home with its location, its history and their friends in the area, Kay says: "Different areas of the house are connected with particular areas of the county or even the particular barn it came from." Specific parts of the home recall not just the family's history, but also the region's. "For example, this sign I'm looking at came from a particular person's house who was the land commissioner for years and years. We have tons of reminders," she says. The family also incorporated many antiques and mementos of their own family life. A childhood wagon was converted to a coffee table. A blue cabinet inherited from Kay's grandmother graces the cabin's living space, and a ladder built by Kay's grandfather leads to a sleeping loft.

Adam loves how they converted community trash piles into a beautiful home: "Some of these boards came from piles of rubble that I drove by every day. Not only did we build the house we live in, but we also cleaned up the town we live in." The family also saved some of the beautiful architectural elements in their region that would have otherwise been lost. "It's like living in New York and one day you're able to put the Empire State Building in your living room," Adam says. "Our house became part of our town, not just another addition."

Build Small

If patience, knowledge, effort and time were the biggest requirements of the Baker family as they built their home, being reasonably sized was the biggest requirement in their home's design. Building small helped keep the project manageable, and Guy and Kay both enjoy living in the cozy space. "I planned to build small because I knew the smaller it was, the easier and quicker I could do it," Guy explains. "I'm glad we did, because the boys are gone now, and it's just Kay and me. We enjoy the small quarters. It's easy to maintain, the power bills are lower, it costs less money. Everything is manageable."

Guy and Kay have always seen the value of living in small quarters. Most of the homes they've lived in have been small, and Kay says it suits her and her family: "We've always laughed over the years. If you were mad at someone, you were just mad, and you better work it out because there's nowhere to run. The idea of going into a room and slamming the door and no one coming in just wasn't in the cards. It's always been a blessing."

Adam says he learned the value of simple living by seeing the excess in the design of area homes over the years. He feels sorry for people who can't recognize the value of a home that's built just for living, not for ornament. "Being that my father was in the construction business all of my life, I've witnessed people arranging houses with designs that include rooms that serve absolutely no function other than looking nice. It's such a shame how those people must live in those houses. There is so much wasted space. If we could only use less space and be more efficient in the design of our homes, it would be half the battle of excess waste."

The Bakers kept the main home small by incorporating several outdoor areas, including an outdoor dining room serviced by this utility shed.

In some ways, the home became a museum of the historical items Guy had collected over the years, thanks to his fascination with antique architectural salvage. Kay sees the home almost as a living museum, where Guy could finally display all of his pieces of the area's building history. "Over the years, whenever Guy worked somewhere and people found out he likes old stuff, they'd just bring him stuff," she says. "He collected things over the years, and it all has lots of memories. It's all in one place now. Anywhere we look or where we're sitting, there's a particular memory."

In the community, the Baker cabin has become something of a celebrity. The home attracts visitors from far and wide, all of whom are blown away by the simple home's soothing feeling. "I had some people come by this morning," Kay relates. "The first thing I told the ladies was, 'There are no female amenities here.' There's no dishwasher, for example. It's a slow, easy lifestyle. We have a wonderful life, but it doesn't really revolve around money." Guy and the boys' clients frequently visit their home. The clients often become so intrigued by the home, they bring other visitors by to see it, as well. One of Guy's clients brought over his sister from Chicago and his mother from Jordan. "People from all over come by. They take a liking to our little home. They just come up in the driveway at will," she says.

Guy and the boys bring the experience they gained building their home to all of their new projects. Folks all around the area have heard of the unique cabin, and they ask for parts of its character to be incorporated into their own homes. Mostly, Kay believes clients hope to capture the feeling of calm the cabin conveys, explaining, "It has a very homey feel, and a lot of simplicity. I think that's part of it, but it's also just the calmness. I don't really know how to explain it. It's just peaceful. You're not trying to make everything perfect. It is what it is and everybody's comfortable. I guess it's contagious when people come in and see and feel that."

The Baker's handbuilt paradise has gotten a lot of attention in their community, and it's driven a good amount of work for Guy's construction business. "We remodeled an old store in downtown Roanoke," Guys says. "Downtown Roanoke is probably 150 years old, and the owner of this building was an attorney who sought me out because of my own home. We used a bunch of old tin because she loved that look." He's happy that the techniques he demonstrated in his own home have carried over into his professional life. And he loves the aspect of historical discovery that comes along with building with used materials: "Over the years, I've been fortunate to not only build new construction, but also to do this stuff we really enjoy. This is fun. You discover and find out so much. One of the simple little things we found when we were using this old lumber was a musket ball from the Civil War era. Someone shot a musket ball into the board or the tree. When we were cutting an old board, we found it."

Building their own home taught the boys that building a home is totally accessible. Jeffery and Kyle have not only chosen to make building their profession, they got started on their own homes at a young age. Guy owns a few rental properties (self-employed, he considers them his retirement fund), and the boys each took one for their own home. "Jeffery, my oldest, had taken a rental house we had built 15 years ago, which is a few hundred yards through the woods from this cabin. At age 18, he remodeled it and put his own thoughts and ideas into it. It's a beautiful home. It's a small cabin. He's been there probably six years now and loves it," Guy says. "Kyle lives in another rental in Roanoke, Alabama. Jeffery and Kyle are both full-time employees working for me. Their interest in building is really high." All three boys are interested in building their own homes. "I would certainly love for me and my two brothers to build a cabin in the woods somewhere," Adam says. "I wouldn't even care if it was mine to live in or not. I'd just love to do another project like that with my brothers. After our project, I had the idea to use broken bathroom tiles and create a mosaic of a Coca-Cola emblem as a countertop. That would be something I'd like to do in my own home one day." Building their family home has instilled all of Guy and Kay's sons with the building bug. "They've all got that desire," Guy says. "They've realized how easily you can do it, for a minimal cost."

Along with connecting with each other and their region, the hand built home also allowed the Bakers to connect with the natural beauty that surrounds them. Their lot, which includes forest and several streams, was valuable to the family for its natural beauty, so they cleared minimal trees to make space for the cabin. The outdoor porches are surrounded by trees. Wildlife wanders through the area. Guy says the connection with nature helps enhance his home's paradise feel. He also honored the trees he did clear by using them in his home's design. Indoor railings are made of twisted pine from trees on the site. Guy designed a bed for the master bedroom — an upstairs sleeping loft — out of local wood, modeled after an expensive bed Kay had admired in a magazine. Sometimes the materials they used led to imperfections, but they make Guy and Kay love their home even more. For example, Guy used wood from trees on the property to make the boards for the floor of the bedroom loft. The fresh wood was still moist when Guy laid it, so he butted them very close together, but they still shrank more than he expected. The resulting cracks in the floor are one of Kay's favorite elements of the home, she admits: "They lend the bedroom a hayloft feel that reminds me of *Little House on the Prairie.*"

CHAPTER 2

A BRIEF HISTORY
OF HOUSING FINANCE

To move forward in housing finance,
we must first examine
how we got to the present.

A S WE CONSIDER the sustainability of housing while we move into the future, we must include financial sustainability in the discussion. In many ways, the current housing market suffers when home values go down. Having a housing finance system dependent on continuously increasing home values isn't sustainable as populations of low-income people in need of quality housing grow. The subprime mortgage crisis was a symptom of the problems inherent in this system. Reduced home values and increasing adjustable interest rates have caused millions of foreclosures. In the second quarter of 2010, 11 million American homeowners were "underwater" (they owe more on their homes than they're worth), according to real estate analytics firm CoreLogic.[1] Rather than suggesting that home values need to go back up, these effects are symptoms of a bigger problem: a system that's dependent on ever-increasing home values most Americans can't afford. If we look with a broader lens, we can see that it would be better for almost everyone if housing prices were to go down. Reduced home prices would allow more people to afford homes. It would bring more tax revenue to cities. It would reduce the amount of our incomes we must dedicate to housing, freeing up resources that could go toward higher-quality food and health care.

■ History of Housing

With any endeavor, as we look toward the future, it's best to first examine the past. Before we can understand the current housing finance system, we need to understand how it came to be.

Government and industry leaders have manipulated American land and housing pricing since the nation's founding. In the very early years, distribution of government lands was chaotic and arbitrary. Boundaries, marked off by footsteps from geographic landmarks, were a common source of disputes. The Land Ordinance of 1785 formalized border markings by astronomical points and broke townships into 640-acre land parcels. Potential homesteaders were minimally required to purchase an entire 640-acre block of land at $1 an acre, just over $8,000 today, expensive during a time when housing finance didn't exist. By 1800, minimum lot size was halved to 320 acres, land prices rose to $1.25 an acre, and homeowners could pay in four installments, according to the National Archive,[2] but forces were still working against low-income citizens who hoped to obtain land. In the South, large-scale farmers became wealthier as crop prices increased, and they bought up large swaths of land at the expense of low-income Americans. The already-settled Northeast held few opportunities, so potential landowners looked toward the growing West. Before the 1846 to 1848 war with Mexico, settlers in the West demanded pre-emption — the ability to settle first and pay later, one of the earliest forms of American credit. After the war, growing numbers of immigrants pressured the government to increase the affordability of unsettled property through homesteading legislation, but corporate interest opposed it; in the North, factory owners feared that cheap land would draw away its low-wage immigrant labor force. In the South, wealthy farmers feared rapid Western settlement would include large numbers of small farmers who would oppose slavery. According to the National Archives, three times — in 1852, 1854 and 1859 — the House of Representatives passed homestead legislation, but on each occasion, the Senate defeated the measure. In 1860, a homestead bill providing Federal land grants to Western settlers was passed by Congress only to be vetoed by President Buchanan. Finally, in 1862 the Homestead Act passed — in large part because the South's secession took the slavery issue out of the equation.

The Homestead Act allowed any citizen who had never borne arms against the US government to attain ownership rights to 160 acres of land by filling out an application, living on the property for five years, building a structure 12-by-14 feet or larger and growing crops. Initially, meeting the requirements was more difficult than it sounded. With no form of mass transportation or outposts for supplies, travels westward through the American wilderness were extremely difficult. Once settlers arrived at settlements on the Plains, the lack of large trees made finding building materials difficult (and it's why many Plains settlers built sod houses). Winds, blizzards and insects made growing crops difficult; limited crops made raising livestock difficult. As a result, many homeowners couldn't stay in one place long enough to make the five-year requirement of the Homestead Act. When the Railroad Act of 1869 passed, relief came to settlements in the form of a nationwide railroad, providing transportation and access to resources. Pioneering flourished. By 1934, over 1.6 million homestead applications were processed, and more than 270 million acres — 10 percent of all US lands—passed into the hands of individuals, according to the National Archives. The Homestead Act wasn't repealed until passage of the Federal Land Policy and Management Act of 1976.

As homesteaders settled the West in the late 1800s and early 1900s, immigrants and others in urban areas were renting apartments or tenements. In growing American urban centers, the need to provide housing for rapidly increasing immigrant populations demanded the construction of thousands of tenement buildings, which allowed property owners to earn a second income renting out low-income housing. These units generally included a street-level store topped by four to six stories that housed four tenements each. For many years, tenements did not have plumbing, and usually several families lived together. In 1890, a studied tenement block housed 4,000 persons in 600 apartments. In 1865, New York City had about 15,000 tenements. In 1900, the city boasted more than 80,000 tenements and 3.4 million residents.[3]

Throughout this period, mortgages existed, but borrowers often had to pay around 50 percent down on five-year loans.[4] While handmade housing continued to be accessible outside urban areas, a variety of forces made purchasing and financing a home created by professional companies more appealing. A more specialized economy gave individuals and families higher incomes and less time and desire to do traditionally homemade, handmade activities such as sewing clothes, building furniture, baking bread and building homes. As the Industrial Revolution took a greater hold, the idea that something was not homemade gained a glamorous appeal. The novelty of factory-made items made them intriguing. Industrialization made its way into the do-it-yourself mindset of early 1900s Americans via hybrid house kits. Sears, Roebuck and Co. famously offered a DIY house kit by catalog, selling more than 100,000 homes from 1908 to 1940.[5] The kits contained more than 30,000 parts and the building instructions for around $2,500.

In the early 1900s, extremely poor housing conditions led to the formation of three organizations tasked with the creation and enforcement of building codes: the Building Officials and Code Administrators (BOCA), serving the eastern and Midwestern states; the International Conference of Building Officials, serving the western states; and the Standard Building Code Congress International (SBCCI), serving the South,[6] but city slums still offered extremely poor housing conditions. Government officials did little to provide decent inner-city housing to the ever-increasing numbers of low-income urban dwellers. "By contrast," writes Anthony J. Badger in *The New Deal: The Depression Years, 1933–1940*, "governments in both Britain and Germany in the 1920s had built a million homes and the Netherlands housed one-fifth of their population in government-owned housing."[7]

■ Moving Toward Modern Housing

The groundwork for much of modern home financing was laid with the New Deal, legislation passed in 1934 by Franklin Roosevelt in response to the Great Depression. The New Deal aimed for economic recovery, reform of economic policies and relief for the nation's jobless.

As the effects of the Great Depression set in nationwide, homelessness began rising dramatically. Hundreds of thousands of homes and farms were foreclosed or in danger of foreclosure. From 1929 to 1933, foreclosure rates rose steadily, peaking at 1,000 a day in March 1933.[8] As rural farmers lost homes, they came to the cities in search of work. Roosevelt's administration hoped to rein in foreclosures and provide low-income housing, and the New Deal enacted three major housing-related programs to work toward those goals.

Home Owners' Loan Corporation

The Home Owners' Loan Corporation (HOLC) was tasked with putting a halt to foreclosures and reversing ever-increasing foreclosure rates. The HOLC used capital funds to purchase and refinance risky mortgages, aiding more than a million homeowners from 1933 to 1935. The HOLC made mortgages more manageable by extending loan periods from then-standard 5-year mortgages to 20- and 25-year loans. In 1935, the HOLC ceased lending, as was mandated in the Home Owners' Loan Corporation Act that created it. By most accounts, it was a successful program, refinancing one of every five mortgaged homes at the time, and lending about $3.5 billion over its lifetime. In addition to mortgage refinancing, its services included debt counseling, budget planning and family meetings. One major criticism of the program is that its policies encouraged racial segregation.

Federal Housing Administration

The Federal Housing Administration (FHA) was established under the National Housing Act of 1934 to insure loans for the construction and purchase of homes. In the 1920s, construction of about 900,000 new homes started each year; in contrast, in 1933, the construction industry started on only 93,000 new homes. By insuring private lenders, the government provided the confidence that lenders needed to offer more individuals the opportunity to build or buy a home. "FHA support encouraged lenders to reduce the down payments they required, lengthen the repayment period, and lower the interest rates they charged," Badger writes.[9] Over the next 40 years, the percentage of American homeowners rose by almost a third.

Greenbelt Programme

The New Deal administration made efforts to create government-built low-income housing communities, though none met with great success. The Greenbelt Programme was designed to put the jobless to work building towns just outside major cities. The communities would offer low-income housing and relieve growing pressure on crowded inner cities. Led by the head of the government's Resettlement Administration, collectivist economist Rex Tugwell, the communities were planned to include democratic decision-making on the part of residents. Many problems plagued the program, however. Its usefulness was reduced by poor site selection near fairly well-functioning urban centers rather than the most blighted. High costs associated with building quality housing quickly made rent too high for city slum-dwellers. Several other programs were launched to attempt to provide high-quality, low-income urban dwellings, but conservative opposition kept successes to a minimum. Attempts at creating government-funded urban housing dropped off until 1949. A post-World War II housing shortage yielded Congressional approval for 810,000 units of public housing, but lack of enthusiasm from both voters and government representatives meant only about 357,000 were actually built. Badger writes:

> Low-cost housing, therefore, never attained public legitimacy and met only a small part of the poor's housing needs. Instead, public housing, in contrast to European countries, served further to isolate and stigmatize the poor, creating the end result that New Deal welfare policies had expressly set out to prevent.[10]

In 1938, the government founded and sponsored mortgage giant Fannie Mae (the Federal National Mortgage Association), tasked with improving accessibility to loans, encouraging lenders to increase availability of mortgage funds and provide liquidity in the mortgage market. Fannie Mae works with lenders to provide mortgage funds. It does not lend directly to homeowners, but purchases mortgages from lenders, holding them in a portfolio and providing lenders with liquid assets to fund more mortgages. The company also issues mortgage-backed securities (MBSs) in exchange for pools of mortgages from lenders. With $2 trillion worth of business, Fannie Mae is the nation's largest provider of mortgages, the second-largest corporation and one of the world's largest financial services corporations. Upon its founding as a government entity, Fannie Mae was only authorized to purchase FHA-insured mortgages.

In the late 1960s, housing went through another set of updates and alterations. Poor residential living conditions and lack of affordable housing led to the addition of the Department of Housing and Urban Development (HUD), passed by Lyndon B. Johnson in 1965 with the mission of creating strong, inclusive communities and affordable housing for more Americans. In 1968, Fannie Mae became a private shareholder-owned company. In 1970, it was authorized to purchase conventional mortgages. Also in 1970, Congress chartered Freddie Mac to help secure low-income mortgages by purchasing and financing mortgages.

In 1977, the Community Reinvestment Act (CRA) was designed to help fight discriminatory lending habits known as redlining. The federal law requires banks and savings and loan associations to offer credit throughout their entire market area. Its purpose is to provide credit, including homeownership opportunities and small-business loans, to underserved populations. Under the law, banking institutions are evaluated to see if they've met community needs and are issued a CRA-compliance rating. Bank CRA records are taken into account when the government considers an institution's application for deposit facilities, including mergers and acquisitions. The CRA has been controversial since its inception, and, as we'll discuss more in the next chapter, the controversy has increased since the subprime mortgage crisis.

■ The Cost of Housing

A commonly accepted general rule today is that we should spend approximately 30 percent of our household income on housing. But why do we accept this figure? What could we do with the extra income we would save if our housing required 20 percent of our income — or 10?

The 30 percent estimate hasn't always been the norm. The number has evolved over time since its founding, based on the US National Housing Act, a Great Depression-inspired public housing program passed in 1937 as a way to provide affordable housing to the nation's lowest-income families. Initial rent limits stated that renters couldn't qualify as "in need" if they earned more than six times the cost of the rent.[11] In 1940, the standards were altered, requiring that rents be capped at 20 percent of household income. In 1969, the standard was raised to 25 percent, and in 1981, it was increased again to 30 percent. Over time, spending greater proportions of our income on our homes has become more acceptable. In fact, we frequently spend more than 30 percent of our incomes on housing. The 2006 American Community Survey[12] found that 46 percent of residents nationwide pay 30 percent or more of their income on housing costs. Thirty-seven percent of owners with mortgages and 16 percent of owners without mortgages spend 30 percent or more of their income on housing costs. The census bureau refers to "30 percent or more of income spent on housing costs" as "housing-cost burden." Families on the lowest rungs of the economy suffer most from expensive housing. While some high-income households may choose to spend more than 30 percent of their incomes on lavish housing and still have plenty of available income, for low-income families, spending more than 30 percent of income on housing cuts into expendable income for quality food and health and child care.

Building homes using the vast supply of used building materials lowers housing costs for the individuals building that home. As several examples in this book show, we can create low- and no-debt homes using almost entirely salvaged materials. If we could institutionalize the reuse of this supply stock, it could have a wider economic effect. Building more homes that cost less to create will reduce the price of overall housing. Though conventionally seen as bad for the real estate market, an overall reduction in the price of homes would mean people would have more disposable income to invest in quality education, health care, nutritious food and home improvements and to reinvest in the economy.

CHAPTER 3

NO-DEBT NEWLYWED DREAM HOME

An Idaho newlyweds' family and friends come together
to camp out in a tent village, scour area dumps,
mix earthen plaster and stack straw bales — and build
the couple's mortgage-free dream home.

WHEN MEGHAN AND AARON POWERS took on building their dream home in 2006, they had a lot going for them. Aaron is a professional builder, and Meghan is an architect (she was an intern at the time). They also have some amazing friends and family members who were willing to travel to Idaho to help build the home, camped out in a makeshift village of tents and a tipi on Aaron and Meghan's land. But still, the couple says anyone could do what they did. It requires hours of research and planning, a connection with your community and a thoughtful consideration of what one really needs in a home.

In 2003, Aaron was intent on building a small home for himself using locally sourced materials. As a first step, he purchased a tract of land in Idaho's Teton Valley. When he and Meghan met and started dating, they began tossing around ideas for the place, incorporating their shared desires to build small, reduce their environmental footprint and live debt-free. "Eventually, the end goal became coming up with a house we were happy with that was mortgage-free," Meghan says.

Their first consideration, both economically and environmentally, was to reduce their home's resource needs by building small. They knew they wanted a comfortable home they could live in for years to come, but they also knew they could create that home in a small footprint by thinking creatively. They started mapping out potential floor plans on a friend's basement floor, drawing out walls and doorways and imagining how the spaces would flow into one another. They spent an entire winter going over the possibilities, walking around in their friend's basement and reconfiguring the virtual house to come up with the best use of space.

The couple also had aspirations regarding their building method. Aaron had long been fascinated by straw bale building, and he wanted to experiment with the building method, which enabled the couple to use materials local to the Idaho countryside. "I have a cousin back in Vermont, and 20 years ago he talked about building a straw structure there," Aaron says. "As a kid, I thought that would be interesting." In Aaron and Meghan's location on the

Meghan and Aaron Powers built their small, handbuilt dream home using straw bale building and materials they salvaged from all over their region.

Betsy Morrison

Idaho-Wyoming border near Jackson, Wyoming, straw bale building is fairly common, thanks to its excellent insulative properties and the thick-walled, cozy homes it creates. "There's a pretty good community of people who have done straw in the Teton Valley. I did a few tours of straw bale homes, and it kind of sealed the deal," Aaron relates.

Having determined their building technique and with their floor plans beginning to take shape, Aaron and Meghan were ready to start collecting materials. A fortuitous trip to their local dump made them aware of how many items they could collect to use in their home. Meghan says they went to drop something off and were amazed when they found more to return home with than they'd taken: "It was amazing the great things we found. Doors, chairs, windows, lumber, drills." They realized using salvaged materials would help keep costs down and reduce their need for new sup-plies. Policies in their area encouraged them. Dumps can be legally scavenged there, though many mu-nicipal dumps have outlawed the practice. Meghan and Aaron say some landfills and waste-processing centers are starting to realize the value of rescuing building supplies before they head to the landfill. "Here, they're trying to divert the waste stream and offer building materials at the landfill, getting them reused in the community. If you have enough con-struction going on, it is a value to the community to not have to dispose of that stuff," Aaron says.

Meghan says that, in their progressive region, local dumps are modeling ways of reuse that are both environmentally and economically viable. Sending construction waste to landfills incurs a fee. When waste collection facilities can find ways to reuse building supplies (often large heavy items) locally, landfill fees are reduced. In this way, the en-vironmentally conscious option also becomes the

The bedroom's clay walls, sliding salvaged barn door and corrugated metal ceiling create a unique, comforting bedroom.

Betsy Morrison

financially beneficial one. "In this area, a lot of our dumps are moving toward becoming transfer stations where they transfer waste, rather than dumping it. It becomes more financially viable to the community to actually use this stuff than to pay a landfill fee," Meghan states.

The large amounts of construction and deconstruction in nearby Jackson also helped make building supplies readily available. Aaron says luck was on their side: "There was a huge house in Jackson, and we got access to the materials from it just before they demolished it." The home — a 10,000-square-foot home being totally deconstructed — provided a majority of the materials Aaron and Meghan used, including appliances, redwood lumber, floor tiles, doors, windows, a barbecue grill and an as-yet-unused elevator, all of which were otherwise destined for the landfill.

Once Meghan and Aaron got the word out in their region that they were on the hunt for reused building supplies, offers started rolling in. Contractors called to let them know of quality supplies that had nowhere to go but the landfill. "In Jackson, we became known as the people who wanted things that were coming out of housing. We often had the ability ourselves or knew people who had the ability to get these things," Meghan says. But Aaron points out that it's not necessary to be "in the business" to get the inside scoop on salvaged materials. The bigger issue was simply getting the word out that they wanted it. People were happy to oblige when the young couple approached them asking for demolition freebies. People's natural instinct is to not want good materials to go to waste. In fact, the offers kept rolling in long after their home was done, Meghan says.

Straw bale building creates thick, cozy, well-insulated walls with deep windowsills.

Betsy Morrison

"Even after our house was finished, we had contractors calling and saying, 'We're pulling these cabinets. They're great. We don't want to throw them away!'"

Although luck was often on their side, the going wasn't always easy. Sometimes Meghan and Aaron had to hunt intensely for supplies in their region, and they had to overcome several minor setbacks along the ways. Meghan's sister, Kathleen Hanson, who, along with Aaron's father, Bob, and sister, Krista, lived with the couple during the summer of construction, says the two stayed calm even when the project got difficult. "One night Krista and I came home from

Building Well

Meghan and Aaron built with straw bales and incorporated passive solar elements into their home's design. Designed to moderate temperatures, straw bale construction's thick walls hold in heat during winter, making them more efficient and cozy than conventional homes. During summer, the thick walls retain evening and early morning coolness, naturally releasing it during the later day's heat. In all seasons, the thick thermally massive walls help moderate temperatures, improving efficiency.

Passive solar design means siting and designing a building to work with the sun's natural heat. For example, incorporating lots of south-facing windows allows the sun's warmth to enter the home during winter and heat floors and windows. Adding shades and awnings keeps summer sun out. Passive solar design requires thermally massive walls and floors, making it a perfect partner for straw bale construction. Meghan and Aaron used a stud wall insulated with recycled denim batts on the south-facing wall, which allowed them to incorporate more glazing.

In Meghan and Aaron's region, straw bale building also offers environmental benefits. "The straw is all local," Aaron says. Straw bale walls are also healthy, because the material is naturally breathable. "Should you have to use some glues, which we tried to minimize, this house doesn't allow it to get trapped inside. It's breathable. You're getting more air exchange through the walls," he says. They used natural earthen plasters both inside and outside the straw bales, which also allows air and moisture to move through the walls naturally. The exterior plaster is made of lime and sand, while the interior is made of earth mixed with sand and straw. The bales are corseted together with bamboo, a rapidly renewable resource. High operable windows allow for whole-house ventilation. A pellet stove and an adobe floor with radiant heat keep the home cozy during the long Idaho winters. The couple also used low-VOC paints, stains and plasters and wheatboard and sunflower board sealed with linseed and tung oil for shelving and cabinetry.

work, and Meg and Aaron were both sitting on top of the straw bales with their heads down," Kathleen remembers. "They had already gotten the straw and stacked it, and they did what they could to keep it weatherized. We were framing and the roof was on, and we were ready for straw. But it was spring, and we'd had a bunch of bizarre weather, and all the straw was rotten. There's all kind of farming and agriculture that goes on in this region, but it was early for cutting, so there wasn't much straw available." She says the crew didn't let their loss of a crucial building supply get them down. They regrouped and started over. "The next day, we all loaded up in the truck and started driving around searching for straw. Krista and I would knock on the neighbors' doors and ask if they had any straw they would sell," she laughs. The crew was just able to collect enough local straw to replace the rotten stock, and the project was back on track.

Hand-poured concrete countertops complement a salvaged faucet, metal ceiling and reclaimed wood beams in the Powers' kitchen.

■ Recycling Renegades

Meghan says that, though recycling has always been important to both her and Aaron, building their home with recycled materials was eye-opening in terms of seeing what a big impact they could make by reusing building materials. Even the few items the couple purchased, such as denim and cellulose insulation, are made of recycled or reclaimed materials. The two know where nearly every material in their home came from. The roof and ceilings are made of metal from Montana sheds. The bathroom tile is a past kitchen countertop — they chipped off the tiles by hand, one by one. Their deep windowsills are made with beetle-kill pine a friend found in the dump. Every piece of lumber and plywood in the home is reclaimed. Meghan recalls: "Perhaps one of the best stories of recycling that summer happened to our sisters, who lived with and helped us for the summer. They salvaged all the electrical wire they could find in one particular house that was torn down. After stripping the plastic sheath off, they loaded the wire in the dump truck, and my sister headed to the steel-recycling center in Idaho Falls. She said that every guy in the shop had to come out to see the 105-pound girl with a truck full of 1,600 pounds of clean copper wire. She left there with about $1,800."

Kathleen also recalls the copper project as among her favorite stories of working on the house, and says it highlighted the foolishness of sending great materials to the landfill. "I walked out with just under $2,000 in cash. We were loving it. It was all going into the dump. It really drove home the point for us as to how much waste we generate building new houses."

Aaron says learning about your community, and tapping the areas where high-quality homes are being torn down, is a first step toward acquiring reclaimed building supplies. He recommends builders hoping to take on a project like this look around to see where people are tearing down old structures to build new ones, and tap those resources. He realizes that he and Meghan had an advantage simply because they live near the wealthy town of Jackson, where vacationers buy lots to build their dream cabin and tear down whatever is already there. "People just need to get out there and let contractors in the community know what they want. It helps to have a community next door where there's a lot of wealth. They're tearing down houses built in the '80s and '90s."

The connections Aaron and Meghan made while building their home — and contractors' desire not to send good supplies to the landfill — carried on long after they'd completed their project. A few years after finishing their own home, the couple helped Meghan's parents build a vacation home in Montana, and they used their connections again for supplies. The amount of money they were able to save was astounding, says Aaron: "We got our hands on a bunch of high-end Pella windows and doors out of a remodel. The local building supply company called me and said, 'This contractor would like to get rid of this stuff,' so we got all these windows and doors from 1999. Glazing is good for easily 30 years, so we used them in that new house. That was $20,000 worth of windows and doors we got for free."

One of the Powers' clever space-saving tricks was to hide a sunken dining table beneath removable floor planks in their living room. When dining alone, Meghan and Aaron typically eat at the kitchen bar. When they want a formal dining space, they use the cozy sunken table.

Betsy Morrison

■ Wise-size Design

Once they had collected all their supplies, Aaron and Meghan started to put together their definitive plans, choosing how to use which material and where, keeping in mind their goal of using space-saving techniques to make a small but very livable home. Much of the design was determined, or at least inspired, by the supplies they were able to obtain. "We ended up redesigning a lot of the house based on the materials we had," Meghan says. "You have to be flexible. Once we got all these different materials, we took inventory and put it together like a jigsaw puzzle."

The interesting, varied materials and their desire for a small, manageable space made for a gorgeous and unique outcome to this particular puzzle. Aaron and Meghan solved many small-space conundrums by using inventive space-saving techniques and dual-function strategies. "A couple of space-saving things are the showpieces everybody talks about," Meghan says. It's easy to see why: The innovative design solutions are striking and seem like obvious, logical ways to maximize utility. "They're unique and you aren't going to find them anywhere else," she says.

Planning Ahead

Though they built small, Meghan and Aaron Powers knew they were building a home to last a lifetime. Aaron says one of the most important elements to building one's own home successfully is "knowing what you need in a home." But he encourages potential builders to consider future needs along with current ones: "If you grow your family and income, you can grow your home." He and Meghan designed their home with two potential additions in mind, in case they needed them for children, extended family or evolving business needs. By considering the right-sized home for them now and later, they hope to have prevented the need to move as life circumstances change.

Aaron also recommends thinking forward when it comes to technology and the environment: "Think about your utilities now and in the future. Even if you [build your own home] on a shoestring budget, you should probably put some of the infrastructure in that will allow you to go to renewable energy in the future." He adds that you don't have to spend the extra money on installing alternative energy when you build the home, but by considering potential upgrades in the future, you can eliminate the need for costly reconfigurations to accommodate systems. "Even if it's just incorporating a pipe for future solar thermal into your HVAC, it's about thinking forward. Maybe you don't have the money to do it now, but try to plan ahead."

Aaron used redwood timbers salvaged from a Jackson home's sauna to build a beautiful round shower. Its exterior is so lovely that it functions as the exterior of the bathroom, eliminating the need for further dividing walls. What's more, the ingenious shower has a removable floor that hides a touch of luxury. "It goes three feet down to be a Japanese soaking tub," Meghan says.

Another innovation inspired by Japanese homes is a sunken dining table tucked beneath a removable panel in the living room floor. The table provides a formal dining table for the occasional times the couple wants one, but it doesn't require the excess square footage a full dining room would need. The dining space is made more special by the couple's connection with the materials they used to build it. "Usually if it's just the two of us, we eat at the bar in the kitchen. If we have guests over, we can pull the floor panel and sit at a built-in table made of cherry cut years ago on Aaron's grandfather's land," Meghan says.

The floor of the shower pulls out to reveal a sunken Japanese soaking tub.

The beautiful redwood for the shower was salvaged from a home in Jackson, Wyoming.

The couple also incorporated many less-flashy ways of saving space. One important component to comfortable small-space design is to include plenty of storage spaces, and to use spots that are often dead space in typical homes, such as high on walls or under stairs. Aaron and Meghan incorporated shelving on interior ledges atop walls that separate spaces without reaching all the way to the home's tall ceilings. Here, they can place books, small storage containers and decor without requiring additional furniture or shelving. The couple used varying ceiling heights to help define spaces and create an open feeling in the home. Spaces with very high ceilings feel bigger, while lower ceiling heights can define more intimate spaces. "Our ceiling is vaulted, and we make it seem bigger by dropping the ceiling planes on the internal spaces such as the closets," Meghan says.

Another key to expanding the feel of small spaces is to create long views. "Viewlines are really important in a small space," Meghan says. "If you are looking outside rather than at a wall, it seems bigger." When designing a small space, it's important to create as many places as possible where views span the entire length of the home. Whether you are looking across one room or across three rooms, seeing a wall that's far away makes a structure seem larger than if it is broken up into smaller rooms. Aaron and Meghan's home is essentially an open rectangle, with spaces defined by partial walls and varied ceiling heights. The open floor plan allows for many views that lead outside. When you look through a home and see the horizon outside as the farthest point, it makes the interior space feel more expansive.

The Powers flouted conventional home design "must-haves" such as multiple bathrooms. They designed their home for their own lifestyle, including lots of storage nooks and beautiful, high-quality materials.

Betsy Morrison

Food from the couple's extensive garden is stashed on high shelves.

■ Hosts with the Most

The outdoors can be a more integral part of a home than just as a visual aid. Designed well, outdoor spaces can become a functional part of the home and expand the living space. As nature lovers, Aaron and Meghan included many outdoor spaces in their home's design, including an outdoor kitchen and barbecue space. The couple also met tons of new friends during their home-building project, and they wanted to be able to continue their community get-togethers after construction was complete. The outdoor spaces allow them to host larger groups of friends and neighbors than might be comfortable in their small indoor space. "We try to throw parties outside," Aaron says. "In the winter, it's more intimate, but in the summer, we'll have 20 or 30 people over."

Along with their invited guests, Meghan and Aaron also frequently find themselves with guests who aren't necessarily friends, families and neighbors. Their home has become famous in the region for its beauty, originality and space-saving techniques. People show up on their doorstep hoping to take a peek. Meghan says most people who come to tour the home love its size and find it as totally livable as she and Aaron do: "I think people don't really have a great grasp of scale sometimes. They think, 'I need 3½ bathrooms and 10-foot ceilings.' But if you ask them to actually measure a ceiling they like, it's usually closer to 8½ feet. A lot of times scale is skewed in peoples' minds." She believes seeing how functional and beautiful a smaller home can be helps people see that big homes don't have to be the standard, and that, in fact, in many ways smaller homes feel more comfortable and natural. She loves that their home is easy to clean and maintain and inexpensive to power and heat. "Society has told us we need certain things that we're often happier without," she says.

The couple says realtors are quick to tell clients that they need three bed-rooms and two bathrooms, or they'll be eliminating future potential buyers. But Aaron and Meghan say they weren't building a home for the next people who would live in it — they were building a home for themselves and the life-style they wanted. "I feel that it's your home and you should do it the way you want," Aaron says. He thinks homes designed with intention will always find eager future residents. "There's going to be someone else out there who is similar-minded and will appreciate what you've done. Maybe that's a good thing. If you've got a cookie-cutter house, it's not going to stand out."

■ Mortgage-free Living

Meghan and Aaron not only love their home for its depth of character, they also love all the things that not having a mortgage has afforded them. "Bigger costs more," Aaron says. Regardless of the number of bedrooms, when one builds a home with no debt, the concept of a "good resale value" takes on a whole new meaning. This couple doesn't have to worry about the housing market. Their small, personalized home is theirs, no strings attached, and that

A neighbor's unused grain bin was put to good use, serving quadruple-duty as Meghan's office, Aaron's workshop, a guest house and sailboat storage.

allows them to focus on other areas rather than worrying about paying the mortgage. "Not having a mortgage opens up doors for you, whether it's not having to work as many hours or having the freedom to pursue your own small business," Aaron says. "The less you're tied to large debt, the more freedom you have to make other choices."

Both Meghan and Aaron have been able to successfully run their own businesses out of their home. Without a mortgage, they had more free capital to invest in starting and running their own companies. Meghan is a self-employed architect and illustrator, while Aaron started Powers Excavation in 2008, and occasionally does freelance carpentry.

Knowing they wanted to work from home also required some design creativity when it came to building their home. Meghan and Aaron both needed a workspace, but they didn't want to add square footage on to their main residence. They discovered an easy answer in an old unused grain bin they found on a local farmer's land. They asked if they could buy it, and their neighbor happily obliged. They built a two-story office and workshop out of the silo. In true Powers form, the silo does quadruple duty, serving as Meghan and Aaron's office and workshop, a guest house for friends and family and a storage garage for the couple's sailboat. Putting the extra rooms in a separate building means they don't have to heat the space on weekends or after hours when guests aren't staying. It also separates the couple's private life from their professional ones. And though their home is small, they didn't want to eliminate luxury. A second outbuilding made of cordwood and an earthen roof houses a small sauna.

The couple's friends helped Aaron and Meghan plaster walls, denail lumber, stack straw bales and much more.

■ A Family Affair

One of the most crucial elements that allowed Aaron and Meghan to build their debt-free home was low-cost labor. But it wasn't just their own. The couple's whole family pitched in. Camped in an array of tents, RVs and even a tipi, Meghan's and Aaron's siblings, friends and parents gathered to help them construct their dream home. "Our process made me feel like it was an old-fashioned barn raising where a community comes together to put something up. It's a day to work hard but

A Building Tradition

Part of what motivated Meghan Powers to want to build her own home was likely her family history. Meghan's parents built their own home in Montana, where they raised their three daughters. Kathleen says their parents extended their can-do attitude to their daughters at young ages: "Growing up, we were super fortunate in the sense that our parents had the attitude, if you want something you can make it yourself." The family's home was handbuilt and a constant work in progress. She recalls that it wasn't until the children were older that they replaced plywood floors for something more permanent, and as small children, the girls were allowed to color with crayons on the temporary flooring. Growing up in such a DIY environment helped the Hanson daughters feel confident in taking on big tasks. "I'm by no means a carpenter, but we'd done a decent amount of building and crafting, so it wasn't totally foreign to me," Kathleen says.

Much as her sister did, Kathleen says she is excited to carry her family's traditions into her own life: "I would love to someday build my own house. It's not something I've done yet, but it's something I always assumed I'd do. My parents built their own house; Meg built her own. For her thesis project, Meg built a little house on my parents' property, and we've been slowly building on that house.

That's something we've all been working on together as a family. It seems like we've always been doing this our whole lives." The idea of building a home is so naturally ingrained in her that Kathleen and her husband have already started collecting materials to follow in her parents' and sister's footsteps. She explains: "We have this shed where we're living now, and we have a couple of pallets of stacked bamboo hardwood flooring that we pulled out of a house in Salt Lake City. We've already started collecting. We have a sink and some cabinets. We're well on our way. My parents have a pottery studio up where we live in Montana, and my mom and I built the sink in Meg and Aaron's house, and I have another sink just like it. We've got all kinds of random things that will someday get plugged into another house." Kathleen loves how little bits and pieces of their various projects weave through her and her family's lives. On the large home she helped Meghan and Aaron pull materials from in Jackson, there were nine chimneys, all capped with copper chimney caps. She remembers standing on Aaron's shoulders and pulling them off. "I have some, and a couple of other friends got some. We have some at the lake house. We all have a piece of the Jackson house. It's just fun," she says.

also to come together with neighbors and friends. We would have a barbecue and a few beers at the end of the day. It was a way to connect," Aaron says.

A core group of five was onsite all summer: Meghan and Aaron, Aaron's father, Bob, and his sister, Krista, and Meghan's sister, Kathleen. Kathleen says the crew had a great time in their summer camp-like set-up: "Meg and Aaron were living out of an old motorhome, and so was Bob. Krista was living in a tipi, and I was living in my dad's hunting tent. They made a cordwood and concrete round structure, which was the sauna house eventually, and we put a toilet in there and we had a shower that hung free. We had this huge mound of stuff we were sorting through and using. It was a fun, awesome set-up. Bob had made this big bermed firepit area where we cooked. We all had full-time jobs, but in the evenings and on weekends, we'd work on the house."

During the six-month building process, other family members and friends came and went, taking part in any and every step along the way, staying in the makeshift tent and tipi camp, "for as long as they could stand," Meghan jokes. Her mother, a potter, constructed a bathroom sink. Friends spent days mudding walls. Krista and Kathleen chipped tiles off the wall of a Jackson kitchen then laid them as flooring in the bathroom.

Aaron says the system worked well for them. "It was a good experience. We both had our sisters onsite. Meghan's parents came down, as did my father. Several friends helped us a bunch," he says. The project also appealed to many people in the area who wanted to try straw bale building and learn about making their own homes. "One thing about straw bale building is that people get excited about it. Anyone can do it," he says. Using a simple building method allowed participation from people of any skill level. "You don't feel intimidated. You don't have a framing gun in hand. It's empowering — at least that was our experience. It allowed a lot of people who didn't have skills in the building profession to get involved and be valuable help," he says.

Though a few tasks required specialized skill, much of the work could literally be done by anyone. Meghan reiterates: "We had a lot of mundane labor. We used old wood, which meant we had to denail, scrape paint…it was mundane help. It was great we had a family that was willing to come out and help us." Like her husband, Meghan says the fact that the project was fun and unique helped encourage her family to come out and help: "It was a pretty unique vision, and they were excited to be on board with it."

Kathleen says several of the construction weekends took on a neighborhood party vibe: "Once we started stacking the straw, there were friends constantly coming and going. Meg and Aaron had a handful of close friends that played a key role in the building. The valley we live in is a fun, hippie sort of place. There were people thinking about straw bale building that had never done it, so there were lots of folks who would come out and help for a day. There was one guy who was a brewer, so he'd make beer and come over, and we'd drink beer and stack straw. We made tons of friends and met lots of people. It was impressive how many people came out and helped."

Meghan and Aaron specifically chose straw bale building because they knew doing a lot of the labor themselves was key to maintaining their debt-free

A Home That Works for Her

In many positive ways, the lessons Meghan and Aaron learned from their unique home cross over into their professional lives.

After working well together on the building project, Meghan and her sister, Kathleen, decided to work together permanently, starting an architectural and botanical illustration business in the grain silo workspace. Meghan also replicates her work on her own home in her architecture projects: "I use it in the design aspect, in both the passive solar and the scale. Finding the right size is probably most important to me." Much as she and Aaron did before building their home, she encourages her clients to first consider how much space they truly need. She often asks clients to measure rooms they like, and they frequently find the spaces are often smaller than expected. She also works to develop creative open floor plans that minimize hallways and underutilized spaces.

Meghan also loves working on projects in which clients actively seek salvaged building materials. Adept at revising building and design plans to accommodate unusual, interesting building supplies, she is happy to work flexibly with her clients throughout the design and construction of a project: "I'm doing a home now where the homeowner is actively searching out ReStores and secondhand. We have an agreement that, as they buy stuff, they send me the dimensions. We just make sure the design can accommodate these things." Though she is happy to help create a home like her own for someone else, she says the hours of determining what type of home works for you and the weeks of collecting materials is in the hands of the homeowners. "I mean, the amount of work we did on our house you can really only do yourself," she suggests, "but I have clients interested."

goal. Straw bale homes are expensive for those who hire professionals, because they are labor-intensive and only some specialized builders use the unique method. But it's also easy to do and requires very inexpensive materials, meaning if you're willing to put in the time to research, learn and build it, you can greatly reduce your costs. "Straw bale isn't cheap for most people," Aaron admits. "But if you supply all the labor, it is very cheap." The couple also had to stay flexible and learn as they went along. Part of the building process was experimentation. Kathleen remembers their trial-and-error method of mixing the mud for the walls: "It was a lot of reading and experimenting. When we were mudding the walls, we were doing test batches to see what would hold and what wouldn't. That part felt like uncharted territory to me, but it wasn't overly intimidating."

Aaron sees putting in the hours of labor as a simple matter of substituting resources they did have for those they didn't. As young newlyweds, they didn't have much in the way of financial resources, but they were blessed with high levels of energy and enthusiasm. "When you're young, you're rich in energy and time, though you haven't hit your peak earning potential," he says. All the work and effort were worth it, the couple says, because they now have a dream home with no debt. Aaron and Meghan value their home more because so much of their friends' and family members' time and effort went into it. Yet Meghan reiterates that, despite all the time and labor, no one stopped their regular lives in order to complete the project: "All of us working on the house had full time jobs in addition to working on this project."

Meghan and Aaron admit that building one's own home may not be for everyone. But any homebuilder can incorporate elements of their project into their own homes. They offer this advice to anyone hoping to tackle a similar project: "Hire a contractor that's willing to let you work beside or beneath him, or hire a friend that's a carpenter. To whatever extent you want to, hold the reins. You don't have to know everything. You just have to surround yourself with the right people. If that can be your friends and your family, fantastic."

CHAPTER 4

THE ECONOMICS
OF RECLAIMED HOUSES

The 2006 to 2008 subprime mortgage crisis
is strong evidence of the need to renovate our housing
finance system, and reusing building materials
is one tool we can use to help make housing more
affordable for everyone.

OUR CURRENT HOUSING MARKET is dependent on continuously increasing home prices for its success. Since the 1970s, it's also a commodity open to speculation. The 2006 to 2008 subprime mortgage crisis — and the ongoing instability of the US housing market — shows that the system is unsustainable. The commoditization of our houses drives everyone from builders to bankers to use cheap materials and inflate prices. Homeowners and investors hope for ever-increasing home values because they view domiciles as financial investments. By using free and low-cost reclaimed building supplies, homeowners are creating zero- and low-debt homes for themselves, and project organizers are creating affordable housing for low-income residents. Creating homes that are built and financed with less reliance on conventional large mortgages could help shift our housing finance in a direction that allows for our homes to require a smaller portion of our financial resources, allowing more of our incomes to be allocated to other things such as better-quality food and health care. An examination of the many factors that led to the subprime mortgage crisis makes its own case for housing finance reform.

Many studies of the housing crash have been published, and many continue to be conducted and published. So many working parts were involved in the crisis that there is no way to pinpoint one cause. What we do know is that many average Americans lost their homes. And that we have no idea what will happen to housing prices in the future. Robert Shiller, a leading economist, professor and writer, told *Business Insider* Editor in Chief Henry Blodget in late January 2011, "It's possible we could launch into another housing bubble if people think that the recovery is real and they want to get in early. We know one thing: According to the Michigan survey on consumer sentiment, people think that prices are low and that good buys are available. There is the beginning of bubble thinking, right there; all it takes is some sense that it's going now. I'm sorry to be so weak as a forecaster. I think it could go either way." By creating our own lower-cost homes with small or no mortgages, we keep our money out of the risky, unstable speculative housing market.

■ History of the Housing Crash

In January 2011, the Financial Crisis Inquiry Commission released a report, more than 500 pages long, on the causes of the crisis. It cites lack of government regulation, failures of corporate governance, excessive borrowing, risky investment and more. Despite the complex factors that worked in concert to lead to the crisis, the results were simple: the near-collapse of our financial system and the investment of trillions of taxpayer dollars into some of our largest financial institutions.

Though I'm not an economist, I believe anyone can take a look at this crisis and see the many flaws that led up to it. In 2006, housing prices were high and had been climbing steadily since the late 1990s. It was considered common knowledge — you probably heard the phrase tossed around — that housing prices could never go down. Investing in real estate was a sure thing, a positive for everyone, because it was the one investment practically guaranteed to return at high rates. And investments in housing weren't returning just 1 or 2 percent, like many investments, but 5, 10 and 15 percent— numbers that made investors hungry to get a piece of the subprime mortgage pie.

It wasn't just investors that wanted in on the action. As regular people saw and heard of huge returns in the housing sector, more and more wanted in. Home-flipping television shows crowded airwaves. In a 2005 study conducted by Robert Shiller and fellow economist Karl Case, a poll of San Francisco homebuyers revealed their economic expectations for their new homes: The mean expected price increase was 14 percent a year; about a third of the home-buyers reported extravagant expectations of up to 50 percent a year.[1]

Economists and government officials encouraged the booming housing market. The government for many decades had attempted to continuously increase homeownership. In a speech in Atlanta in 2002, President George W. Bush stated, "Too many American families, too many minorities, do not own a home." Though American income levels stagnated during the eight-year Bush presidency, housing prices soared. In an interview, Bush's first Treasury Secretary John Snow said, "The Bush administration took a lot of pride that homeownership had reached historic highs. But what we forgot in the process was that it has to be done in the context of people being able to afford their house. We now realize there was a high cost," reports the *New York Times*.[2] As the authors of the *Times* article, titled "Bush Drive for Home Ownership Fueled Housing Crisis," Jo Becker et al., say, the housing market was one bright spot in a difficult economic climate that prioritized cutting taxes and privatizing Social Security. "Ever-rising home values kept the economy humming, as owners drew down on their equity to buy consumer goods and pack their children off to college."

In cities across the nation, housing prices rose drastically from the early 2000s to the mid- to late 2000s, by 30 to 50 percent and more. Although a study conducted by Shiller and published in *Irrational Exuberance* shows housing prices in the US had stayed fairly constant with relation to inflation and population for the 100 years before the 1990s, housing prices jumped by 30 to 50 percent and more in the 10 years after 1998. In London, a home that cost around £100,000 in 1983 and £110,000 in 1995 was valued at about £330,000 in 2008. A Boston home that sold for around $100,000 in 1985 and around $110,000 in 1995 was worth about $225,000 in 2008.[3]

Though it seems clear in hindsight that something was going awry, even leading economists at the time seemed unaware, or unwilling to report, that housing prices were out of whack. According to the *Times* article,[4] Lawrence Lindsay, Bush's first chief economic adviser, said there was little impetus to raise alarms about the proliferation of easy credit that was helping Bush meet housing goals: "No one wanted to stop that bubble. It would have conflicted with the president's own policies."

As early as the middle of the Clinton administration, government economists issued warnings that the housing bubble may burst. Shiller writes that, in 1996, government economists and policy makers called him and his colleague John Y. Campbell to testify before the Federal Reserve Board about their beliefs that the housing bubble was going to burst. Shiller says, though he and Campbell expressed their concerns that the bubble was going to cause severe damage to the economy, no government policy changed.

In fact, George W. Bush increased the homeownership incentives that had been put into place by President Bill Clinton. Bush said, because of languishing incomes and increasing home prices, he would "use the mighty muscle of the federal government" to meet his goal of increased homeownership. The *Times* article says government policies at the time also encouraged the formation of small mortgage brokers and required continually lowering loan standards. "Bush pushed to allow first-time buyers to qualify for government-insured mortgages with no money down," Becker et al. write. And though "Republican congressional leaders and some housing advocates balked, arguing that homeowners with no stake in their investments would be more prone to walk away," the president was committed to increasing rates of homeownership for low-income residents — in theory, a good idea for social welfare, but not when low-income residents were taking out loans they couldn't afford to pay back. The president leaned on mortgage brokers and lenders to devise their own innovations, the *Times* writes. Bush said, "Corporate America has a responsibility to work to make America a compassionate place," according to the article. Bush also insisted government-backed lending giants Fannie Mae and Freddie Mac meet ambitious new goals for low-income lending, encouraging provisions that allowed first-time buyers to qualify for government-insured mortgages with no money down.

■ Market Forces Behind the Bubble

The major driving force behind the subprime mortgage crisis was the fact that homebuyers were taking out loans they couldn't afford — though it wasn't all their fault. While there was some irresponsible borrowing, there were also a number of factors working together to lead homebuyers to take out high and risky mortgages. And while government regulators seemed to rely on free market to discourage home loans to individuals who couldn't afford it — after all, lenders are invested in getting their money back, so they require a reasonable amount of evidence (such as income level and credit history) to assure them homeowners could afford to repay loans — in this case, lenders were eager to put just about any applicant into a loan because they were not the ones responsible for collecting on them. The loans were being bundled and sold up a chain to Wall Street, and eventually on to high-level investors. Rather than having to wait to get their loan back with interest over 30 years, loan originators could get a quick payout by selling the promise of hundreds of mortgages-worth of money.

Lenders, driven to sell more mortgages, told potential homeowners they could afford the investment because their home equity would grow so fast, they could either refinance their loans based on the increase in equity they would see in a short time, or they could sell the home for a profit. By the mid-2000s, homebuyers with no credit history or income verification were being approved for NINA — No Income, No Assets — loans. Mortgage brokers didn't care if homeowners ever paid their mortgages back because a greater-than-ever-in-history pool of investors was chomping at the bit to buy up low-income mortgages.

Some of the following information relies on the National Public Radio (NPR) series *This American Life* program #355, *The Giant Pool of Money*. The "global pool of money" refers to the amount of money held by everyone in the world. In 2000, the global pool of money was valued at around $36 trillion.

It had taken hundreds of years to accumulate to that quantity. Six short years later, the global pool of money had grown to $70 trillion, largely because formerly poor countries had increased their wealth. This meant there was roughly twice as much money in the world that was seeking profitable investment.

Over the same period, from 2000 to 2006, the US Federal Reserve, under Chair Alan Greenspan, was keeping interest rates at very low levels (around one percent), meaning global investors couldn't make money on US treasury bonds. Looking for stable investments with a decent rate of return, the investors turned to US mortgages. They contacted big investing firms like Morgan Stanley and asked to get in on the mortgage market. Mortgages at the time were returning five, seven and nine percent interest as the number of American homeowners and the value of American homes both grew. But these international investors couldn't buy single mortgages, and they didn't want to be involved at that level, so they turned to mortgage-backed securities — investments on large numbers of home mortgages bundled together.

So a chain developed that connected low-income subprime mortgage applicants with Wall Street investors: Small mortgage brokers sprouted up all over the country, offering loans to low- and no-income applicants. As pressure mounted from both the government and the financial sector, and deregulations allowed for fewer and fewer barriers to buying loans, a perfect storm was created.

Mike Francis was "executive director at Morgan Stanley on the residential mortgage trading desk at the beginning of the implosion," he said in an interview for *The Giant Pool of Money*.[5] He said this about mortgage bonds: "It was unbelievable. We almost couldn't produce enough [bonds] to keep the appetite of the investors happy. More people wanted bonds than we could actually produce. That was our difficult task, trying to produce enough. They would call and ask 'Do you have any more fixed rate? What have you got? What's coming?' From our standpoint, it's like, there's a guy out there with a lot of money. We gotta find a way to be his sole provider of bonds to fill his appetite. And his appetite's massive."

To increase mortgage-backed securities (the bonds the investors wanted), Mike Francis had to buy more mortgages. He asked colleagues like Mike Garner, the head of the largest private mortgage bank in Nevada at the time, Silver State Mortgage, for more mortgages. Garner located individual mortgages and bundled them to sell to Mike Francis. It's interesting to note that Garner had literally zero prior experience in the housing finance market — he'd left a bartending job to start his pop-up mortgage lending business.

As interest in bonds skyrocketed and regulation all but disappeared, lenders began employing all sorts of creative techniques to attract new potential homebuyers "with a proliferation of too-good-to-be-true teaser rates and interest-only loans that were sold to investors in a loosely regulated environment," writes Becker et al.[6] Predatory lenders also specifically targeted minority homebuyers. A study by two Princeton researchers published in the *American Sociological Review* in October 2010 shows that minority homebuyers were disproportionately targeted for subprime mortgages, and experienced a disproportionately high level of foreclosure following the crash, compared with white borrowers.[7]

As investors continued seeing good returns on mortgage-backed securities, the need for ever-increasing numbers of mortgages exceeded the number of qualified borrowers. Lending standards were continuously dropped to allow more people to qualify for home mortgages. Individual mortgage lenders felt so much desire from investors to buy up home loans that they had virtually no risk in any mortgage they financed. Restrictions got looser and looser as corporations told government regulators such as the Security and Exchange Commission that the housing bubble was based on real value and was a boon to everyone in the US — including homeowners.

Meanwhile, Bush populated the financial system's oversight agencies with people who, like him, wanted fewer rules, not more. Bush's first chairman of the Securities and Exchange Commission promised a "kinder, gentler" agency, according to the *Times* article:

> Bush's banking regulators once brandished a chain saw over a 9,000-page pile of regulations as they promised to ease burdens on the industry. When states tried to use consumer protection laws to crack down on predatory lending, the comptroller of the currency blocked

the effort, asserting that states had no authority over national banks....
The president did push rules aimed at requiring lenders to explain loan
terms more clearly. But the White House shelved them in 2004, after
industry-friendly members of Congress threatened to block confirma-
tion of his new housing secretary. In the 2004 election cycle, mortgage
bankers and brokers poured nearly $847,000 into Bush's re-election
campaign, more than triple their contributions in 2000, according to
the non-partisan Center for Responsive Politics.

As the number of homeowners and value of loans increased, investors were
buying up bonds based on speculation prices that assumed the people being
approved for loans could afford to pay them back. All the easy money fueled
its own cycle. Low-income citizens heard it was easy to get a mortgage, buy a
house, sell it in a year and make some cash. They thought they could finance
adjustable rate mortgages, pay only the small introductory interest rates, then
move the loan along to the next buyer. That same thinking pervaded all the
way up to the investment brokers on Wall Street.

As we know now, the bubble burst, and the entire system collapsed.
The economics of the housing market proved unsustainable. Eventually,
large investment firms had to settle their books, and as increasing numbers
of mortgage-owners foreclosed, the bottom dropped out. Borrowers of the
popular ARM (adjustable-rate mortgage) loans (about 80 percent of subprime
loans) weren't able to refinance as the borrowers had planned, and once the
introductory rate was over, interest rates soared to double and triple initial
rates. Investors lost money; people lost homes. In 2008, home foreclosures
were up 225 percent compared with 2006, and 2008 saw more than 3.1 million
foreclosure filings.[8]

The value of the houses was no longer based in reality — the market had ar-
tificially inflated the prices of homes to far above their actual value. Borrowers
were saddled with debt two and three times the real value of their homes.
And, though loan regulations have been increased greatly since the collapse,
we've done nothing to fix the fundamental problem with our current housing
finance system: It treats homes as for-profit commodities, and it is designed

to only be successful if housing prices remain stable, or go up, but not if they go down.

■ Reducing the Cost of Housing

Housing prices today are still elevated by inefficiencies and high salaries. Overall housing prices can, and should, go down. One way we could create more affordable homes is through the use of all the materials we already have access to in current buildings. Though it is better for current investors if housing prices go up, in the long run, it is better for our society if housing prices go down. According to an economics principle known as Baumol's Law, housing prices should, in fact, go down over time because they are within the category of "goods and services whose production is amenable to technological progress."[9]

But Shiller says the concept of ever-increasing home prices is supported by several widespread myths that have grown up around housing and that help convince us that it's logical for housing prices to continually increase — "the myth that, because of population growth and economic growth, and with the limited land resources available, the price of real estate must inevitably trend strongly upward through time."[10]

Shiller uses the term "new era stories" to describe cultural myths that are not based in facts but that over time become widely believed because they are spread by word of mouth, focused upon by the media and falsely appear to be supported by unrelated factors. He says three "new era stories" support a belief in ever-increasing housing prices.

Myth: Land Is Limited

The first of the new era stories that contribute to our ability to believe housing prices should never fall is the idea that land is scarce. The argument goes like this: Housing prices can never go down because, even if more houses are built, we are running out of available land. Therefore limited land will lead to rising land prices, keeping housing prices on a continual upswing. However, Shiller counters, land is not limited. In fact, "according to the US Census for the year 2000, urban land area accounts for only 2.6 percent of total land area in the

US. The high value of homes in major cities is accounted for by their location to the built environment, not the unique value of land."[11]

Although we should prize undeveloped land for its many benefits, land as a commodity is not limited. If we reconsider how and where we build our urban centers, we can create built environments that work with, rather than endanger, the biological systems around them. Though we must use our land resources with great care, a scarcity of land is not, in economic terms, justification for ever-increasing home costs. Expensive home prices that exclude low-income homeowners won't lead to less need for land to be developed. Wisely designed cities that incorporate healthy spaces for low-income residents would be the best way to keep land development at a minimum.

Myth: Building Materials Are Limited

The next new era story that supports the myth that housing prices should continually increase is a supposed explosion in construction labor and material costs. The myth is that our building materials are scarce, and that their future reduced availability will drive home prices up.

Not only are our current building supplies not limited, literally hundreds of alternative building materials and methods exist, many of which are examined in this book. As the projects in this book show, we waste tons of building materials annually. Our current quantity of recycled building materials alone could supply much of our future needs. Each year in the US, we demolish approximately 250,000 buildings, according to the National Renewable Energy Laboratory,[12] and we build about 150,000 new buildings each year.[13] Along with our vast stock of reclaimed housing supplies, and those available in existing buildings that will be modified in the future, many other alternative materials can also be used to build homes, as explored in chapter 7. Homes can be constructed from straw, clay, mud, aluminum cans, recycled tires and more.

Even if we limit our conversation to our current typical building materials — lumber, plaster, concrete, glass and steel—a supply shortage wouldn't support ever-increasing home prices.[14] Lumber is a renewable resource. Well-managed commercial forests can be run efficiently and effectively to continuously produce this resource. Advances in our cultivation, management and use of quicker-growing wood-producing plants such as bamboo and wood-replacement plants such as hemp will also help reduce our need for slow-growing trees.

Our other building materials are available in vast quantities, as well. Shiller explains:

> Gypsum, the main ingredient of plaster and wallboard, is a very common mineral. The White Sands National Monument in New Mexico, famed for its views of massive white gypsum-sand dunes as far as the eye can see, contains 275 square miles of nearly pure gypsum sand. Although this particular site is protected, it is worth noting that White Sands alone would be enough to supply the world's construction industry for hundreds of years. Limestone, the principal ingredient of the cement used for concrete, represents approximately 10 percent of all sedimentary rock formations on earth. Glass is made primarily, and sometimes entirely, from quartz. Quartz is the second most common mineral in the earth's crust. Iron, the major ingredient of steel, is the fourth most abundant element on the earth's surface, constituting 5 percent of the surface.[15]

Today's most common building materials are not in limited supply, and as we will see in many examples, homes can also be constructed from straw, clay, mud, aluminum cans, recycled tires and much, much more. The notion that home prices will continuously increase based on limited building supplies is false.

The Myth of the Glamour City

The third new era story supporting our false belief in continuously increasing home prices is the myth of the glamour city. This new era story posits that the limited amount of space within our nation's most desirable cities means housing prices will continue to increase because people will be clamoring to live in fewer available homes, particularly in desirable city neighborhoods. Although we will likely always see location premiums in great neighborhoods and cities, it doesn't follow that home values nationwide would surge continuously because everyone will sacrifice many other aspects of their quality of life such as health, personal wealth and stress level to live in these areas. Increasing availability of affordable housing in well-designed livable cities and neighborhoods would drive the prices of urban-center housing down. According to the principle of supply and demand, a greater quantity of livable urban areas would drive down the prices of urban housing nationwide. Advances in building technology allow us to create taller, more efficiently and safely designed residential-use buildings than ever before. We will continue to provide increasing residential opportunities in our current cities.

As we will examine more fully in chapter 12, we can also create new urban centers. It often seems to us that our cities, with their generations of history and culture, could never be matched by other locales. However, people enjoy cities for a number of reasons. New urban centers provide citizens with the opportunity to shape their new home; once established, new urban centers develop a character of their own — within 50 years, they offer a rich and developing history. We often forget how recently some cities, once established, have come to be. Environmentally, creating more medium-sized cities is more advantageous than continuing to expand geographically outside current metropolises. Currently the most inefficient and most popular type of city development, suburbs are responsible for more carbon emissions per person than rural areas or urban centers. Most suburbanites rely almost entirely on personal vehicles for transportation. They travel farther distances not only to work, but to accomplish tasks usually performed on foot in the city. Suburban children often go to schools far from their homes, and centers of commerce lie in strip malls accessible almost exclusively by car. More people clustered in smaller areas require fewer resources and have a lower carbon footprint than those who travel long distances to faraway city centers. Increasingly intelligent city design and rising populations will work together to increase the number of attractive cities. Lester R. Brown, president of the Earth Policy Institute, cites one example in his book *Plan B 4.0: Mobilizing to Save Civilization*. Developer Sydney Kitson has acquired 91,000 acres in southern Florida on which he is developing the environmentally and socially minded community Babcock Ranch. The city will be almost entirely powered by an on-site solar photovoltaic energy facility, according to the Babcock Ranch website,[16] and will be designed for pedestrians, mass transportation, community interaction and harmony with nature.

Regardless of any current limitations of residential spaces in our current cities, demand for ever-increasing city living options will drive us to create more and more urban living opportunities, not fewer and fewer as would be required to drive up housing prices. By ensuring we are using our urban spaces wisely and efficiently, and by creating new urban centers, suburbs and towns designed for local community living, we can reduce overall housing prices.

■ The Benefits of Affordable Housing

Despite the fact that a reduction in home values represents a loss for some homeowners and investors, over time, reduced housing prices benefit everyone, freeing our resources for other pursuits. While an overall reduction in the cost of housing may have short-term negative effects on current investors and some homeowners, the long-term economic and health benefits of providing low-income housing outweigh any losses that may be incurred by housing price reductions. Reclaiming our supply of housing materials rather than dumping it all in the landfill could help reduce the expense of providing and subsidizing affordable housing. HousingWorks RI, a coalition of nearly 140 organizations working to ensure that all Rhode Islanders have a quality, affordable home, says that studies have "quantified the economic impact and multiplier effect of investing in affordable homes." The group says an initial Rhode Island investment of $37.5 million into the Building Homes Rhode Island affordable-housing program generated nearly $600 million in total economic activity, a $15.80 return on every $1 invested.[17] Providing affordable housing frees up household income for other uses, improving health and the economy. A study on the positive impacts of affordable housing on health conducted by the Center for Housing Policy[18] found numerous health and economic benefits to providing quality low-income housing. The study lists these nine benefits:

- Affordable housing may improve health outcomes by freeing up family resources for nutritious food and healthcare expenditures.

- By providing families with greater residential stability, affordable housing can reduce stress and related adverse health outcomes.
- Homeownership may contribute to health improvements by fostering greater self-esteem, increased residential stability and an increased sense of security and control over one's physical environment.
- Well-constructed and managed affordable housing developments can reduce health problems associated with poor-quality housing by limiting exposure to allergens, neurotoxins and other dangers.
- Stable, affordable housing may improve health outcomes for the elderly and individuals with chronic illnesses and disabilities, by providing a stable and efficient platform for the ongoing delivery of health care and other necessary services.
- By providing families with access to neighborhoods of opportunity, certain affordable housing strategies can reduce stress, increase access to amenities and generate important health benefits.
- By alleviating crowding, affordable housing can reduce exposure to stressors and infectious disease, leading to improvements in physical and mental health.
- By allowing victims of domestic violence to escape abusive homes, affordable housing can lead to improvements in mental health and physical safety.
- Use of "green building" and "transit-oriented development" strategies can lower exposure to pollutants by improving the energy efficiency of homes and reducing reliance on personal vehicles.

By institutionalizing a method of using reclaimed materials to reduce the cost of building new low-income housing and renovating current blighted housing, we can benefit urban dwellers and our economy. As the subprime housing crisis proves, our current system of housing finance is not sustainable, and we can use homes built of reclaimed materials as a springboard to change the system.

CHAPTER 5

THE ART OF A HOME

A couple outside Boulder, Colorado,
builds an artistic, nature-inspired dream home
with the help of their community.

W HEN NAOMI AND RICK MADDUX got the opportunity to purchase
the land on which their former home was located near Boulder, Colorado, through a Boulder County land-acquisition program, they were inspired
to begin work on their dream home. Building slowly and thoughtfully, and
taking their design inspiration from the natural beauty that surrounds them,
they viewed the project as an art project as much as a home. Thanks to large-
scale community work parties, the home also provided the couple an opportunity to connect with others and to demonstrate an alternative way of building
a home.

A professional woodworker, Rick Maddux says the impulse to create has
been with him since childhood: "I've known forever that I had to make things.
It's a passion to make things, to take things apart. I was the kid who took apart
the clock to see how it worked. My mom would say, 'Could you please put the
toaster oven back together?' She was great about it because she's creative as
well. And now that's what I do for a living."

Naomi Maddux, a singer-songwriter and professional organizer, reveals
how the couple's project changed her: "I became a glass artist in the process
of building this house. I didn't know anything about it before. I loved making

music, but I've never been a visual artist in my life. This house opened me to create in a visually artistic way."

According to Rick and Naomi, they owe the gorgeous results of their giant art project to their collaboration, which for both, pushed their boundaries and stretched their notions of what a home could be. Naomi says the home required a huge contribution from each of them: "I'd say it was the combination of his skills and my capacities, which over time became stronger and stronger. Soon I was able to communicate to Rick my vision, and he would tell me what he thought was possible. This was our primary way of collaborating."

Rick and Naomi Madduxes' handbuilt, straw-bale home was a work of art they built with friends and neighbors over several years, and with no construction loan.

Rick says building the home was the result of his and Naomi's combined passion for building. The two had tried their hand at building on a smaller scale in a previous home, and they reused many of those efforts in this home. But this time, Rick says he really wanted to fulfill his and Naomi's ultimate vision: "The inspiration really just came from a curiosity to actually do something that we had a vision for. We build things together, Naomi and I, as a passion since we've met. We built a house before this one, and a lot of that house is in this house."

Naomi Maddux learned to create stained-glass artwork through building her Colorado home, and now works in stained-glass professionally.

■ A Home Plan

Once they purchased their land, the two knew it was time to build the home they'd live in for the rest of their lives. And they knew they wanted to take their time, design it well and avoid a construction loan. "We made, I'd say, 13 or 14 sketches of ideas of what we wanted and what we needed," Rick says. "We knew we wanted something that would be really interesting. We had everything from kidney shapes to earth ships — all sorts of drawings. We finally agreed on a pretty simple design that included flow and simplicity, and was something we could actually build ourselves, by hand."

Rick and Naomi knew they wanted to build the home by hand, so they decided on a simple structure that is essentially two stacked boxes. They were both interested in straw bale construction, and they knew they wanted to use most of the materials from their previous house. Combining these elements led to their initial house plan. "We'd been very interested in straw bale for quite some time," Naomi says. "In this region, it makes a lot of sense. I love the aesthetics and the materials. And I love doing things with my hands. Straw bale construction is something a layperson can do easily if they have the time

Naomi and Rick wanted their unique, rounded front door to be one of the first elements guests see, rather than a garage door. It makes their home feel welcoming.

Paul Weinrauch, courtesy Boulder County Home & Garden magazine

and inclination." They chose to use reclaimed and salvaged materials for nearly everything that wasn't straw and plaster, Naomi explains, because it felt right for them to incorporate history into the home, and to create a smaller footprint by reusing materials.

The two didn't have to travel far to obtain their first big piece of architectural salvage. It showed up next door. "One of the first things we purchased was our roof material because our neighbors bought too much and we bought the rest from them," Naomi laughs. "It was one of the last things we put on and one of the first things we purchased. We made a shed area to put things we collected along the way." The two took advantage of both their connections and their community by mining the local architectural salvage store. Boulder is fortunate to have a local reclaimed building materials yard, and Rick and Naomi were lucky enough to know someone who works there and could call them with tips. "We have a dear friend — he actually introduced my husband and me — who works at

ReSource. He would often give us a call when he got materials in and give us tips on what was coming in," Naomi says. But that doesn't mean they weren't out hunting for supplies. She adds, "We'd check alleyways and dumpsters and look on the street and go to building sites that were tearing off roofs. We became excellent scavengers. There's the sort of organized scavenging, where a person does it for you, then there's scavenging you do yourself, finding people who are tearing things down and who are happy to have you carry them off for reuse."

As they delved into the long process of collecting building materials and planning construction, they worked to nail down the home's design. They could both identify specific things they wanted and didn't want in a home, and

The Madduxes wanted their home to be unique and imbued with character. One of their design requirements was to avoid hallways.

Paul Weihrauch, courtesy Boulder County Home & Garden magazine

they designed around that wish list. "We wanted it to feel warm and welcoming, so we paid attention to that in our vision," Rick says. "One thing we both don't like is that, in most American houses, the first thing you see is the garage door. So we knew we wanted to put the garage out of sight. We said, 'Let's put it over there on the east side. Let's put it underground and grow a live roof on top.'" The southern-facing slope of the hillside was a perfect invitation for the couple to incorporate passive solar design. By orienting the entire house to the south, they assured all rooms would face the warming winter sun, yet be protected from overly hot summer rays. "We planned a prominent front door and stairway in the center that says to visitors — You are welcome, please come in!"

The pair also knew they wanted the home to feel spacious inside, without actually being a large structure. "If you can see from one side of a living space to the other, even if it's only 30 feet wide, you get a 30-foot view," Rick says. "Our house is 50 feet wide, and there are two places right when you walk in where you can see from one side to the other on the first floor. It feels big even though it's only on a 20-by-50-foot footprint." They used their experiences from past homes and their design sensibilities to establish simple rules based on their preferences. When they put their wish list and design rules together, it defined the home's layout. "We said, 'Let's not have any dead ends. Let's avoid hallways.' And that is how it is. There are no hallways. When you come in you can go left, right or up, and if you go left or right, you can continue all the way around," Rick explains.

The materials they found as they searched for salvaged items also played a major role in determining the final outcome of the project. The interior design was in large part about accommodating objects they found, rather than about finding items to fit spaces they'd created. "That really carried through the whole house," Rick says. "A straw bale is 3 feet wide by 1½ feet tall by 14 inches thick. So you design the house with 6-foot-wide windows that are 3 feet off the ground so you just can pop those four bales in beneath the window without cutting them. That's one part of building that I know because of my profession: The more you fit things in their natural shape, the less you have to alter and force them. It was the shape of the building blocks of the straw bales that finally determined the overall rectangular shape of the house."

■ Building Community

As the design evolved, the time to start the actual construction began. Knowing they wanted to do straw bale, Naomi and Rick started volunteering on other people's straw bale building projects to get a sense of things and learn more about it before they tried it themselves. They used their many connections and friendships in the area to find places where they could learn, and they tapped those same resources when it came time to start organizing their own work parties. Naomi handled most of the organization of the parties, which she says were amazingly productive, as well as incredibly fun and totally hectic: "We had been hosting social gatherings here for a couple of years before doing this project, mostly for full moons, equinoxes and solstices, and so we had an initial base of people I could e-mail saying, 'Hey, do you want to learn straw bale building hands-on? We're doing a work party here.'" She and Rick also became members of the Colorado Straw Bale Association, which allowed them to invite group members to their parties. "It started with our network of dear friends, but it grew quickly," Naomi says. "To date we've had about 100 volunteers working on our house."

Naomi says it's essential for anyone hoping to take on a project like this to develop the skills to organize small-

The simple methodology and safe, non-toxic nature of straw-bale building with earthen plaster walls means friends and neighbors of every age could help at Naomi and Rick's community building events.

scale events, such as scheduling people's time, communicating effectively and planning ahead in terms of supplies and materials. "The trick was to coordinate those work parties, which was no small feat," she admits. "When you have a bunch of people at your home, you have to make sure they know what they're doing and that they have the right materials to work with. You have to be up for that kind of thing." It was important to Naomi that their friends and community members be informed and well-supplied enough to have learned something and be productive, but it was also important that they have a good time. "For instance, we'd have to make sure we had enough plaster mixed ahead of time so they could be moving and laying the plaster. We'd set up stations so three people could work at the same time, so they weren't isolated and could be social and work with each other. It was kind of like spinning plates. You'd start one here, get them all the materials and make sure they have what they need, then you'd move on. You teach enough so the first people can teach the next people who come in, because not everybody would come at the same time. It was a bit of a three-ring circus."

But Naomi knows the project would not have been possible without the huge amount of community participation, and she and Rick were extremely thankful for their community's generosity with time and effort. "A tremendous amount of labor got done that way," she acknowledges. "Carrying the straw was cumbersome. The plastering requires literally tons and tons of material to be lifted and troweled onto the walls. It was a godsend to have that much help. It was amazing how much work got done in one day with so many hands pitching in."

She says the safe, simple and non-toxic nature of straw bale building was perfect for community involvement, and encouraged people to participate: "Even kids could come do it with their bare hands. You don't have to worry about wearing gloves or getting it on your clothes. It's really friendly that way." And though she likens the idea of community involvement to the old barn raising tradition, she says straw bale is an even better project for people to work on together: "You don't have to have much skill at all to do it. As opposed to barn-raising, where carpentry skills are helpful, just about everybody can do earthen plaster. We've had everybody from kids to 70-year-olds coming and helping." She adds that there is one more vital component to having a successful work party: "I think you have to be social. You have to know some people and know how to work with people and appreciate what they do — and you have to feed them! Make sure you have plenty of good food ready so you can just serve it up when folks get hungry — the hard work makes for hungry people!"

Building at a slower-than-typical rate allowed the Madduxes to use artistry in their home, and to test out new methods as they went along. Naomi tried out scrap-tile mosaics in an upstairs bathroom, then repeated the process throughout the home.

Paul Weinrauch; courtesy Boulder County Home & Garden magazine

■ Slow Going

Naomi says building at a slow pace was another requirement that made building a home possible for them. This allowed them to try techniques in practice, rather than planning the entire structure virtually: "When you build something quickly, you have to think about it abstractly. You have to put it all on paper and then finish it, wham bam, get it done. But often what happens on paper is not how it feels within the environment itself." Taking it slow also gave them the time to incorporate all the beautiful materials they were continuously collecting throughout the entire process. "When you take it slow, you can take it room by room, aspect by aspect. You can use the materials you come across. For example, we found a really cool claw-footed tub that just fit into the small area we had designated for a tub, and we were able to reuse the rock walls we built in our previous home's bathroom around it."

Naomi and Rick also took advantage of the ability of testing out building techniques over time. The slow pace gave them the time to learn and experiment with various artistic strategies. When something didn't work, they had time to go back to the drawing board and come up with a different strategy. When something did work, they could replicate it in other areas of the home. "I first tried doing glass mosaic for a countertop in the upstairs bathroom. I'd never seen or done it before, but I had a lot of scraps so I tried it," Naomi explains. "And when I finished, I said, 'Wow, that turned out pretty good! I could do that treatment in different areas of the house!' Building slowly lets you see how something worked in one part and how you could fuss it and use it again in a different application, or in a different way." When a friend donated a Mexican handmade sink for the downstairs bathroom, she was able to create a scrap-glass mosaic countertop that correlated with the design of the sink.

Although they didn't save or finance a large quantity of money to pay for someone else to build their home, they still had to reserve resources in order to work on the project, Naomi says. Working slowly made the project financially feasible without a construction loan. "For some people, it's a matter of collecting enough money to pay somebody to do something you want done. In our case, it was more a willingness to set aside time and be able to not have an income when we were doing building spurts," Naomi explains. "By going slowly, we could afford to do that. If we had done it quickly, we wouldn't have had the funds to pay ourselves and pay for the project at the same time."

For Rick and Naomi, the slow pace also made the project manageable — without going crazy. Though building the home was a big part of their lives during construction, they were conscientious to be sure it didn't dominate them. "You can still have a life and have a working life and keep up your relationships and not make yourself crazy busting your ass to get this thing done," Naomi says. "It has a different cadence to it that's more within the realm of living a balanced life."

■ Reclaiming Character

The couple says using reclaimed materials also gives their home a difference cadence, and a connection to the past. Rick and Naomi love the stories that the historic items bring to their home, making the home more interesting and every part alive with its own story to tell. "My house has a cast-iron sink we got from a music commune," Rick says. "So there's a story right there. This sink has the energetic history of however long it's been in that old farmhouse. It's an old thing. It's heavier-duty than what you could get now. It's a relic. It looks different, and it makes people curious. They ask 'Where did that come from? What is that about?'" He offers another example as he looks at a wooden post along the north wall of

Salvaged materials look and feel different from things available today, Rick says. Their kitchen's heavy iron stove and exposed reclaimed wood boards lend it a cozy, old-world feel.

Paul Weinrauch, courtesy Boulder County Home magazine

the house: "That post's got a nickel stuck in it that I found there when I was building. When that post was a horizontal wall piece, it was by a kitchen table. Somebody stuck it in there and forgot about it. There's that element of 'What's that about? Why is that there?' It's everywhere in this place."

Rick says people recognize the difference between the items in their home and the objects of today. And he and Naomi have stronger connections with their reclaimed items, feeling as though they layer their own story onto its history. "We pulled that heavy cast-iron sink out of that old farmhouse, and we put it in the little house we built before this one, and it lived with us there for a while. Then we needed a new house, and sure, that sink is going in the kitchen. We don't have to shop. We've got the sink, and we'll build the rest of the kitchen around it."

The many historic items imbue Rick and Naomi's home with character. They both feel the old materials bring a good energy with them — a strong sense of lives lived well. "The salvaged materials have an energy of their own," Rick says. "Energy is important in a household. There's the shell, and then there's the soul of the house. That's what people feel when they come. These things aren't bright and shiny and perfect and crisp. They have some history, some experience."

Along with their ability to bring character and history to the home, salvaged materials are higher quality than those one can find today, Naomi says: "I think, once upon a time, craftsmanship was more important than economics. People just did things beautifully, and they made things with really good-quality materials. For the most part, when you use a reclaimed material — depending how old it is — it's a much better quality than what's being produced today." She loves that her home's unique nature is enhanced by the fact that some materials simply don't exist anymore today. No one would see these items in a newly or recently constructed home, and they're beautiful. "Some of our ceiling wood is this old 12-by-1 siding you can't find anymore. It's beautiful cedar, absolutely gorgeous," she enthuses. They know that these high-quality old materials are hardy and sturdy because they've already stood the test of time. They're unique in size and shape, as well as in durability. "Our old oak flooring was reclaimed from an old house in Denver. We used it in the cabin we built before this house. We tore it out and stored it and then used it on the walls here. You can't get planks that are 20 feet long anymore. Today you get them in 2- and 3-foot sections, 4-foot if you're lucky. They have stood up for all this time. They were used on some place, taken out, used another place, taken out. Anything from Home Depot isn't going to stand up to that kind of use and reuse," Naomi says.

Another positive aspect Naomi associates with using high-quality antique materials is their beauty, which allows for the materials themselves to be works of art: "Even today's hemlock you can pick it up, and it floats in your hand. We had 2-by-4s that are Douglas fir. You couldn't carry two at a time, just one at a time. You want to show off the beauty of the wood itself. Most 2-by-4s you want to bury in the walls, but you could turn this into something gorgeous because of the beauty in it."

By using architectural salvage, Rick and Naomi were able to make their building materials part of their home's decor, and essential elements in the art project that is their home. "Our doors are handmade out of 3-by-4 tongue-and-groove cedar planks that came out of a roof on an industrial building. The handle is made out of an antler. The window in the door is a stained-glass piece that's rounded and custom-created for that door," Naomi says.

Rick and Naomi worked together and compromised on every aspect of the home, including where and how to use bold color.

Courtesy Boulder County Home & Garden

■ Living in an Art Project

Naomi says the slow building pace and the interesting materials were both contributing elements that allowed her and Rick to develop their home as a living art project. The two weren't only dedicated to finishing a home in which to live; they were interested in fulfilling their artistic visions. The front door was a key factor in their vision. "The door was created to look as if it's a rounded door," Naomi reveals. "By creating a rounded false-front panel, the door itself could be square and create a tight-fitting jamb. It closes flush to the rounded front panel, so from the outside it appears to be rounded. That door was an idea I had as to how to make something square look round. Then my husband figured out how to do it." This is an example of how the two worked together on the artistic development of the home.

The inspiration for the project came from the couple's innate desires to explore and create, as well as from the materials. Knowing they would have to customize the home to fit its unusual materials anyway, the couple decided they might as well express themselves at the same time. "It has to do with a person's aesthetics and the kind of materials you are working with," Naomi says. "When you work with reclaimed materials, you have to customize them to your situation anyway. If you have a choice to customize something in the usual way or in an artistic way, you might as well do it artistically. It's all about your sensibilities."

As artists, Rick and Naomi Maddux place value in beauty for beauty's sake. Naomi says creating a beautiful work of art was more important to them than finishing the project quickly: "For a lot of people, the main goal is to get the structure up and be done with it. We were more interested in making it as beautiful as we could. It's not about efficiency

Naomi and Rick were inspired by their natural surroundings. They hoped to build a home that fit in with its landscape and had a light impact on the environment.

and economy; it's about doing something from your heart's desire. If your heart's desire points you in the direction of beauty, and you have the time and inclination to do it, then that would be your choice." For their project, Naomi and Rick chose to value beauty and a connection with history and nature over efficiency. Though modern culture often values efficiency and economy above all else, Naomi says she and her husband followed their own set of values as they built their home: "A lot of the signals in our culture point us in the direction of if it's not economic, why bother. But we didn't have a construction loan. We were able to do this as we felt inspired to."

The main inspiration for the couple's art project lay in their property and their sense of connection with the natural world. Naomi and Rick's appreciation for the land around them helped them understand the value in creating something beautiful, rather than something economical. "This land is so beautiful. It was a matter of meeting the beauty of our environment with what we could muster as human beings," Naomi says. "We were inspired by our surroundings. There are gorgeous forests here, and there are wildflowers in summer and a creek that runs through. There's water and the sun shines a lot, and the seasons are really palpable. I think all these things contribute to giving you a sensibility about the magnificence of nature and what value that exquisiteness has unto itself."

In many ways, Rick and Naomi attempted to mimic the natural landscape within their home. "It's a pretty organic house," Rick says. "There's a lot of posts and beams exposed. The hand rails are aspen limbs with river birch banisters." Incorporating his life's passion into his home, Rick integrated woodworking details all around the house. "The main beams are from one tree — it was a dead-standing Ponderosa pine. I cut the tree in half and used the larger base of the tree on one side and carved a 20-foot papa whale on it. On the other side, I used the upper portion of the tree and carved the mama and a little baby whale."

Rick defines art as something that evokes emotion, and he says his home communicates the story of its own creation: "Art is something that makes you feel. You may hate it, you may love it. It's a lot like relationships. In relationships, sometimes it's hard and sometimes it's easy. There's stuff in this house we built when it was really easy, and they look like something. It's not good or bad, but it creates a feeling. Then there are things we built when it was hard for us personally. The art expresses that."

For example, the shower, according to Rick, was an extremely difficult part of the project, one he and Naomi struggled to get through: "The walls in the shower were really hard and challenging. It was really hard to do. That brought out the depths of who we were, living through that experience." But today he appreciates the memory of working through the challenge: "I shower every

All the love and attention, as well as salvaged materials, created a home with a uniquely personal feel, Rick and Naomi say.

day, and I remember that — us getting the walls and how heavy it was and how hard it was."

Rick admits that collaborating on a home and piece of art required compromise, and taught both him and his wife to give and take: "The siding in the kitchen is from this old farmhouse, and you can't get it anymore. It's 12-inch-wide old-growth cedar. I wanted it white. I *really* wanted it white. She *really* wanted to color it. It was the same with the ceiling of the dining room, which is also the floor of the bedroom. She wanted it green. I went with it, and I love it." Working together on the project provided an opportunity for the two to grow in their relationship and as individuals. Rick recalls, "I had to say to myself, 'OK, she's pushing my limits. Go for the color in the kitchen.' And it's OK. If you're involved with a project like this, there's joy and angst. She knows what she wants, and I know what I want. This was a place I could find a compromise and test my limits." The project reached into other areas of Rick's and Naomi's lives, as well. Rick explains, "Now I'm profiting. I sold a dining room set because of its colored seats and back. It's expansive that way."

■ A House Called Home

All the love and history and artistry that went into the house created a home with a spirit of its own. "It's palpable. You can feel it," Naomi says. "Mostly I feel it when I'm away and I come back. Or when we have people over and they say, 'Oh my god, this place feels so good.'" She believes people can feel all the energy that was poured into the project — from the natural materials themselves, her own, her husband's and their whole community's: "I think it's all that loving attention and all those hands. I think it has a phenomenal effect on how the entire place feels. I think you can feel the harmony that's involved. If there's a lot of dissonance in your process, you're going to feel that, too. It depends how it was done and if the people involved were having a good time and enjoying themselves. It comes through the walls."

Rick feels that the quality and history of the materials and the spirit of cooperation are evident to everyone who visits the home. And he loves his home's connection with the land around it: "Everybody finds this place very interesting. It's unique. It's not like a regular house. There's natural plaster on the wall. You use a flour paste as the glue and just really fine clay and really fine sand to make an earthen plaster, and you can add your own pigments. People ask, 'What's that sparkly stuff in the wall?' And we say, 'Oh, that's mica that we found around some quartz up on the hill.'"

Naomi loves that their unique home project has been able to touch and educate so many in their community. First, the project educated the large quantities of people who volunteered. "A lot of people were able to learn skills they can apply to their own projects. They were able to experience how doing something together for the love of it feels. That was a really cool thing," she says.

Now that they've finished the home, the couple continues to host gatherings in it, which allows them to demonstrate their unique building technique to others. Their involvement with the community has led to more outreach possibilities for Rick and Naomi to demonstrate slow building. "I've had permaculture groups come up here and use the site as an example of natural building techniques, craftsmanship and the love of place. This home seems to be an inspiration to others of what is possible, both for kids and adults," Naomi says. "For example, I used to take care of a young girl when she was a small child. Now she's in 4th grade, and her mother had the idea for me to collaborate with the class to create a stained-glass piece for their auction to raise money for the school." After working with the class, Naomi invited them to visit her home and glass studio. "I had a group of 4th graders up here for a site visit. I talked about this kind of building, about doing it this way — with a whole community and with an emphasis on beauty and art. We talked about the value of doing something by hand," Naomi remembers. Naomi believes everyone who visits the home is able to feel the way it's different from a typical home. "The teacher said, 'OK, class go in and don't touch anything.' I said 'they can touch everything! In fact, I'd prefer it if they did.'" Naomi says. "They were incredibly respectful. They had a great time. A lot of the kids said, 'Wow, when I grow up I want to be able to do something like this.' They got that sense of creating something that was more open and inviting, that had the elements of the environment in it, and the freedom to be able to explore. I talked about this kind of building, about doing it this way — with a whole community and with an emphasis on beauty and art. We talked about the value of doing something by hand. I got such incredible feedback from the teachers and the class. At the end, the teacher walked up and said, 'I've never seen them so calm. I think your house has an incredibly calming effect.'"

After all the hours and love they've put in, Naomi and Rick feel their home is an extension of themselves, and it makes it invaluable to them. "We don't plan on ever selling the house. It'll stay in the family one way or another," Rick says. "It's really satisfying to both of us to have the memory of how this old farm door got the stained glass in it. So it's very easy to keep the commitment to keep it in the family. It's not about making it to turn a profit; it never was. That is what happens in conventional construction, and that changes the

■ A Mini-mortgage

Rick and Naomi Maddux didn't take out a construction loan for their home. They have a small mortgage on their land, but their slow and artistic building style wouldn't have lent itself to the usual schedule for repaying a construction loan. Had they relied upon conventional methods, Naomi doubts the home would even exist: "We probably couldn't have qualified for a loan at the time we built. We didn't have enough accumulated income." When the couple started construction ten years ago, banks were uninformed about nonconventional building methods and did not lend to builders whose plans were outside the realm of "normal" building, according to Naomi: "It may be easier now, but at the time, getting a construction loan for a straw bale project was almost impossible. That wasn't a deal for us. We didn't intend to have a construction loan." Naomi knew the constraints of a conventional loan would make the project unmanageable for her and Rick's life, acknowledging, "With a loan, you have to be finished within six months or a year max. There was no way

I wanted to be constrained like that. I didn't want to blow a gasket trying to get everything done in that time." Building quickly would have yielded an entirely different result than the home they ended up with. "It would have been a completely different animal, having someone else call the shots. We weren't into that. We didn't have an end date, and that was just fine with us," she says.

Naomi feels that living with a small mortgage has allowed her and Rick to lead a calm, self-sufficient lifestyle. And she enjoys knowing they are self-sufficient. Their home doesn't rely on the stock market or the status of the housing industry. "I think it's just allowed us to live within our means," she explains. "We are not big spenders. It makes it so we can breathe every month. For the most part, we live on a small budget every month, and we know we can do that and we're not in debt. I call it peace of mind. We don't have this weight over our heads. We're not paying interest. We're not wasting our money paying some capitalistic entity to float our boat. That feels good."

whole flavor of it. We get to live in something we built ourselves. It keeps its value real. It's about heart, humor and humility. It makes it soulful. You can't help it — even if it isn't a piece of art, even if it's not so green — if you build your own house, it feels more like yourself. That's something that's really precious."

CHAPTER 6

BUILDING BASICS

Using reclaimed materials is a great way to reduce
our homes' ecological footprint, and reclaimed materials
can also work as a perfect complement to several
forms of green building. Here, we'll discuss some of the
green building options available to builders and ways
to improve a home's efficiency and the affordability
of its maintenance over time.

CONSIDER YOUR CLIMATE and your wishes and desires before choosing a building system. If you live in an extreme climate, a building method that includes thick, highly insulating walls is a good option. Consider examining your region's vernacular building styles — before the advent of cheap heating and air conditioning, people relied on wise building methods to keep their homes comfortable year-round. Studying this vernacular architecture can give you excellent input on the way to site and design your home to work with your climate. For example, in the South, long, narrow buildings with ventilation on either side encouraged breezes. Large porches shaded interiors, and detached kitchens kept the heat of cooking fires away from the main living space. In the American Southwest, Spain, Greece and Italy, light-colored adobe structures reflected sunlight, keeping interiors cool. In the frigid Northeast, rounded rooftops on Shaker buildings were designed to shed winter snow. On the plains, long sod homes helped insulate and protect interiors from heavy winds.

Two key elements when considering any building style are insulation and thermal mass. Insulation refers to the amount of conditioned air the home keeps out, while thermal mass refers to the walls' ability to store and release heated and cooled air. Thermal mass helps regulate temperatures. For example, in summer, a home with high thermal mass walls will cool down overnight, then release cool air as the space warms in the heat of the day. Thermal mass has the opposite effect in cold temperatures, storing the day's heat and releasing it when nighttime temperatures drop.

■ Earthen Homes

Earth is one of the world's oldest building materials. Several models of earthen homes exist. Their thick earth walls can hold and release heat, but they don't have high insulative properties, making them excellent for warm climates, but not ideal for extreme cold. One of the most recognizable types of an earthen home is the Southwest's adobe structure that is constructed from bricks. A clay mixture is baked (often in the sun) to form these bricks. Another type, "rammed earth" homes, uses forms that are filled with layers of compacted moist mineral soils. When the forms are removed, the walls are able to support loads. A common method of vernacular building, rammed earth was "rediscovered" as an alternative building method in the 1970s and has been standardized and modernized, according to Rammed Earth Works, a California company. Cob building combines clay, sand and straw to make a slurry that is shaped into walls without machinery. Cob allows for creative structural home construction — walls can be formed to be nearly any shape, and cob creates a charming feel reminiscent of Old World cottages. The straw content in cob gives walls a higher level of insulation, but it is still not ideal for extremely cold weather. Cob has been used for centuries, and medieval cob homes still stand in both Europe and the Middle East.

■ Straw Bale

Straw bale building is excellent for cold climates because of its thick, well-insulated walls. In these buildings, straw bales are stacked to form walls, which are then covered with thick earthen plaster. The straw bales offer insulation, while the plaster provides thermal mass. Sometimes, building authorities require straw bale walls to have a conventional post-and-beam supporting structure, with the stacked bales as wall filling. Straw bale associations across the country often host work parties, in which group members come together to stack bales and apply plaster. Participating in work parties can offer a hands-on way to learn the building method. Straw bale walls are dense, and frequently straw bale builders include a "truth window" — a small section where earthen plaster is not added so you can peek at the bales on wall interiors. Straw bale is considered a healthy home option because the straw and clay naturally moderate humidity levels, and the breathable walls don't trap toxins inside. Straw bale is typically not appropriate for extremely humid climates.

■ Concrete Forms

In any discussion of green building methods, you are likely to hear about ICFs (insulated concrete forms) and SIPs (structural-insulated panels). ICFs are rigid foam forms reinforced with steel and filled with concrete. SIPs are panels of oriented strand board that sandwich a piece of foam insulation. Both products provide excellent insulation and durability, though they do not provide thermal mass — the ability to store and release cooled or heated air. It's vital that these efficient methods employ excellent ventilation techniques, because the highly insulated walls aren't breathable. Factory-formed, neither is ideal for the do-it-yourselfer, and they can both be expensive materials, though prices can match those of conventional timber-frame construction due to decreased labor costs, and their extreme efficiency mean these homes can save energy over time.

■ Cordwood

Cordwood building uses short round pieces of wood (like firewood), stacked and mortared into place. The wood provides good insulative properties, while the mortar offers thermal mass. The wood used in cordwood building would otherwise be considered a waste product in the construction industry, as it's not in the form of a board. One can also incorporate other materials into cordwood walls, such as glass panels to admit light. The method is fairly easy to achieve without much training.

■ Timber-framed Construction

German in origin, timber-frame is the ancestor of our conventional building method today, but it uses larger posts that extend from the floor all the way to tall ceilings, and all joints are formed with mortise-and-tenon joinery rather than nails or screws. Generally, the beautiful woodwork in a timber-framed home is left exposed. Though they are beautiful and can be made green using sustainably forested wood, high levels of insulation and alternative energy systems, timber-frame requires an advanced set of building skills. These homes are typically very expensive and very durable.

■ Light-frame Construction

Light-frame construction is the conventional building method used for most homes in the US. Walls are made of wood/particleboard forms filled with insulation. If using reclaimed boards and healthy insulation, light-frame can work well with reclaimed materials. If you use subcontractors, this is the type of building they will be most familiar with. It is also possible to combine light-frame with other building methods, such as cob or straw bale.

■ Earthships

Though earthships aren't a building type per se, they are a design system that incorporates many natural and recycled materials. They follow six main design principles:

- Thermal solar heating and cooling. Earthships rely upon the sun and the Earth's stable temperature to moderate indoor temperatures naturally.
- Solar and wind electricity. Earthships produce their own electricity with a prepackaged photovoltaic/wind power system.[1]
- Contained sewage system. Earthships reuse all household waste in treatment cells that feed food gardens and landscaping.
- Building with natural and recycled materials. The most important quality

in earthship construction is that the main building materials have high thermal mass. Earth, stones, concrete or other thermally massive materials could all be employed. Earthships also use sustainable, indigenous and reclaimed materials. Many are built using reclaimed tires that are filled and tightly packed with earth then stacked.

- Water harvesting. Earthships capture stormwater for use in the home and gardens.
- Food production. All earthships incorporate at-home food production.

■ The Importance of Design

No matter which building method you choose, it is key to sustainability that our homes are efficient and affordable to maintain. Using reclaimed materials reduces a building's environmental footprint in terms of materials used. Making smart choices in design and materials selection can go a long way toward improving a home's affordability and efficiency over time.

If you're designing a home from scratch, the first consideration is site selection and home orientation. Position a home to take advantage of the sun's ability to aid in the heating and cooling of homes, a method known as passive-solar design. South-facing windows capture the sun's heat in winter and direct it to thermally massive interior elements such as concrete and stone, which hold the heat during the day and release it slowly over time. In summer, overhangs block heat gain from the sun, and thermal mass holds cool evening temperatures and releases them during the warmer day. Several building methods that complement the use of reclaimed materials create thermally massive walls. Straw bale is an excellent insulator and provides thick, thermally massive walls. As we'll see in Chapter 7, the Phoenix Commotion achieves thermally massive walls by incorporating reclaimed materials into concrete.

It's also important to consider wind speed and direction and other natural phenomena as you site your home. Well-placed windows can help achieve natural ventilation, reducing the need for air conditioning. Ensuring your home is filled with natural light can help reduce energy use from electric lights. Skylights or high clerestory windows help bring sunshine deep into interiors, as do open floor plans.

Proper insulation levels are essential to an efficient home. Plenty of recycled-content and natural insulation options exist. Energy Star offers guidelines for insulation levels in attics and floors.[2] In new or renovated homes, achieving high insulation levels is key to creating an efficient home.

Windows and doors are also essential elements of a home's efficiency. Reclaimed doors are excellent for efficiency. Generally made of solid sturdy wood, antique doors are often more insulating than modern doors, which are often hollow and made from particleboard. Windows are a bit trickier. Though beautiful, historic windows are usually single-pane and very inefficient. Modern low-emissivity, dual- and triple-pane windows provide much better insulating properties than antique versions. High-quality storm windows can help improve the efficiency of older windows, but they can't achieve the efficiency of new windows.

Be sure to consider efficiency when choosing household appliances, as well. Eschewing a clothes dryer can completely eliminate one source of household energy use. Consider a clothesline (several non-permanent, retractable versions exist) outdoors and/or in the laundry room and a drying rack. For most appliances and fixtures, including clothes washers, refrigerators and toilets, buying a new efficient version is worth its investment in energy savings over time.

Residential sources of renewable energy can reduce maintenance costs over time, especially as unstable energy prices go up in the future. Obtaining low-cost building supplies may help make room in your budget for alternative energy; even if you can't afford it now, consider designing your home to accommodate future alternative energy systems. Simple design tweaks such as making sure wiring is easily accessible can greatly reduce the cost of adding alternative energy systems, as technology improves and systems become more affordable.

Water savings and processing is another key consideration. Many economists predict that major worldwide water shortages will be our next most pressing environmental challenge. Water use also requires energy; transporting water in and out of your home requires energy, and as much of our water use is hot water, decreasing our water use also decreases our energy spent heating it. Building in ways to manage this increasingly precious resource will help reduce a home's maintenance costs and improve its environmental footprint. Several very simple water-saving measures exist: Choose low-flow shower-heads and faucet aerators, along with dual-flush, low-flow or composting toilets. Simple graywater systems use water from faucets, showers and clothes washers to irrigate landscapes. You can also design systems that reuse sink water to flush the toilet.

Your landscaping can also be designed to help process water, reducing the amount of water sent to area treatment plants. Select permeable paving or gravel instead of hardscaping. It allows water to go through plants' natural filtration process and seep into the water table, rather than running off into street gutters and on to treatment facilities. Also consider the grade surrounding your home when designing landscaping. If gutters lead directly to paved areas or the street, you are directing rainwater to sewage systems. Attempt to design your home so rain from gutters flows onto your yard and through a system of plantings, soaking into the ground. You don't want gutters to dump water right next to the house. Grade landscaping down as it moves away from your home to direct rainwater away from foundations. You can also attach rain barrels to downspouts to collect rainwater for landscape irrigation.

Landscape plant selection also influences your home's energy efficiency and resource conservation. Choose native plants that are naturally resistant to local pests and are adapted to your climate to reduce the amount of irrigation they require. You can landscape to enhance your home's efficiency by planting deciduous trees that block summer sun but admit winter sun, planting windbreaks upwind of drafty areas and using plants and trees to shade overly sunny areas and more. You can read more about it in Natural Home & Garden magazine.[3]

As you plan your home design, reduce its environmental footprint by carefully considering how much space you really need. Limit square footage by using smart, multifunctional design and incorporating lots of storage. Well-designed outdoor spaces also can enhance living area without increasing your home's square footage. Long views and high ceilings make homes feel bigger. It's often more environmentally friendly to build up rather than out, limiting the amount of land you develop. If you have spaces you only use occasionally, such as a guest house for children who have moved away or a part-time or seasonal workshop or workspace, you can reduce your main home's footprint by placing those spaces in an outbuilding not attached to the main home.

The health of your home is also an important element of its sustainability over time. No one is going to want to live in a toxic home. Try to reduce home items that include formaldehyde and formaldehyde-containing glues. Particleboard often includes toxic chemicals. Using reclaimed materials brings its own set of health concerns. Reclaimed materials are more likely to be made of natural materials such as solid wood, and they're less likely to be chemically treated. However, reclaimed materials with paint on them are likely to contain lead. Don't attempt to remove lead-based paint. Either paint over it with new zero-VOC paint, or retain its aged look by sealing it with a clear non-toxic sealant (AFM Safecoat makes a few varieties). Carpet often contains synthetic materials and chemical glues; choose natural wool carpet or opt for hard-surface flooring such as reclaimed wood. Carpet can also harbor chemicals tracked in from outdoors.

PART 2

INSTITUTIONALIZING REUSE

CHAPTER 7

A TALL ORDER
IN TEXAS

The Phoenix Commotion's unique, artistic homes
keep waste out of the landfill and communities alive
in Huntsville, Texas.

I N 1998 IN HUNTSVILLE, TEXAS, Dan and Marsha Phillips founded the
Phoenix Commotion, a home-building business dedicated to providing low-
income housing for working folks — mainly single-parent, and low-income
families and working artists. The company keeps costs low by building with
salvaged materials collected from all over the world. Made from an unusual
and diverse assortment of building materials — landfill-bound scrap wood
and insulation, used wine corks, adobe, stone, bottle caps, shards of broken
tiles and mirrors, papier mâché, old plates and CDs, to name a few — Phoenix
Commotion homes don't look like the standard suburban ranch house. These
homes, part perfectly personalized living space, part living art project, look
more like something dreamed up in a fairy tale than something you'd find in
a subdivision. As Dan Phillips explains, a town the size of Huntsville throws
away enough waste to build a small house every week. He's trying to build
those small houses.

■ Phoenix Commotion Basics

Dan Phillips has had several careers throughout his life, and today he applies
skills from all of them to his dream job. Spending ten years as a dance professor at Huntsville's Sam Houston State University (SHSU) gave him skills in

instruction and artistry. Long dedicated to the idea of creative reuse and making new from old, Dan and his wife, Marsha, opened an art and antique restoration business called the Phoenix Workshop in 1985. The store gave him the chance to see all the great "waste" materials coming out of the construction industry in his area. But it wasn't until 1998, after 35 years of marriage, that they decided to make good on Dan's lifelong dream of becoming a builder. They mortgaged their home and founded the Phoenix Commotion, Dan's salvaged-home-building business. By 2003, the Phoenix Commotion had taken off to such a degree that Dan could no longer split his efforts between the Phoenix Workshop and the Phoenix Commotion, so Dan and Marsha closed the Workshop to focus full-time on the Phoenix Commotion.

One of Phoenix Commotion's many unique designs, "The Treehouse" is made of reclaimed lumber.

Named for the mythical phoenix, which continually renews itself from its own ashes, the Phoenix Commotion makes its business out of using resources others consider worthless. Dan builds affordable homes with salvaged, slightly damaged and other unwanted building supplies he obtains for free from manufacturers and wholesalers. He is driven by a twofold commitment: to keep usable building materials out of the landfill and to provide homes for the working poor. The first, he says, was borne of a childhood with parents — a lumberyard-working father and homemaker mother — who had lived through the Great Depression. "We never threw anything away,"

Dan recalls. As a child, he would go to landfills with his parents and search out usable materials. He built his first bicycle at age 14 from parts he found there. He says his fascination with saving old things from their landfill fate found a perfect match when he realized the great need for affordable housing in his area: "I had always suspected that one could build an entire house from what went into the landfill, and, sure enough, it's true. Then, with the crushing need for affordable housing, [building homes] was a natural connection."

The "Storybook House" is one of dozens of Phoenix Commotion homes that are taking over whole sections of the city.

Phoenix Commotion

Along with keeping building supplies out of the landfill, using free materials helps Dan achieve affordability. So does his crew. Every Phoenix Commotion project is manned by a staff of untrained minimum-wage builders, community volunteers and the homes' future owners, all under Dan's expert tutelage. Of his crew, Dan says the building program is part job, part apprenticeship: "I only hire unskilled workers at minimum wage, but they get a fire-hose of information and training during their tenure on the crew. When they have enough skill to compete in the marketplace, I push them out the door when permanent jobs become available at a higher rate." Each untrained, paid crew is occasionally supplemented with a slew of volunteers from the community and SHSU. The final crucial component of Dan's workforce is the future homeowner — he requires every homeowner to participate in the building of her home.

■ Building a Phoenix Commotion Home

The first Phoenix Commotion house was a Victorian-style charmer constructed over 18 months beginning in 1998. The exterior walls contain no studs; rather, they are made of blocks of western red cedar that Dan and his crew stacked, glued and toe-nailed together. Victorian houses often have turrets — small ornamental towers on the front of the home. The Phoenix Commotion Victorian is no different, except that its turret is constructed of stacked, nailed and glued nine-inch chunks of western red cedar 2-by-4s. In an artistic unification of indoors and out, Dan constructed a high-backed master bathtub using the same method. Hickory nuts and egg shells filled with Bondo provide decorative accents on the outside of the house — a homespun alternative to marketed counterparts.

The two main tenets of a Phoenix Commotion home are reuse and efficiency. By gathering materials from all over the world, the company provides a valuable service to its building-supply donors: a guilt-free, non-landfill destination for post-market building materials. Word of the Phoenix Commotion has gotten out to everyone from local businesses to large national ones such as Weyerhaeuser and McCoy's building supply companies. Rather than pay a hefty fee to send their leftover supplies to the landfill, companies ship materials to Dan's warehouse, providing tons of valuable lumber and supplies for free. Because scavenging is banned at Texas landfills, the key is getting access to landfill-bound supplies before they're sent there, Dan explains: "I go to wholesalers, and the scenario is this: The wholesaler will sell 75,000 square feet of tile, and then the person who bought it says, 'I really only need 72,000 square feet.' The wholesaler says, 'OK, fine,' and they end up with the extra 3,000 square feet, which isn't enough to sell at a wholesale level and ends up at the back of a warehouse. Eventually they say, 'Let's get rid of this stuff. Let's give it to Dan.'"

Inside the "Treehouse Studio," frame corner samples from a photography framing store combine to create a zigzag-pattern ceiling.

Because wholesalers must pay a fee to ship materials to the landfill, sending it to the Phoenix Commotion saves them money. The donation is also tax-deductible, making it extremely financially attractive to wholesalers. After a few years of doing this work, Dan says that he rarely finds himself in short supply: "Over the years, the companies know what I'm doing, and they ship it to me. One afternoon, I'll receive an 18-wheeler with $50,000 worth of stuff. Last fall, I got an 18-wheeler of redwood. Two months ago, I got three 18-wheeler-loads of framing lumber — 2-by-6s to 2-by-12s. They say, 'It got a little gray on the end, let's give it to the Dan.'" He estimates getting at least three phone calls weekly with offers of free supplies. "My advice to anyone who wants to take this on: First get a warehouse, then start sniffing around. It'll come by the 18-wheeler."

Without a program like the Phoenix Commotion, companies have literally no non-landfill options for efficient disposal of excess or slightly damaged materials. Wholesalers could theoretically sell them online or to private buyers, but the economic payoff doesn't justify the man-hours required to sell small quantities; they simply won't do it. With no infrastructure in place to reuse these materials, the wholesalers' and manufacturers' options are limited: Dump it in the landfill, which is wasteful and expensive, or store it, which takes up too much valuable space. The Phoenix Commotion provides a better option.

Dan doesn't discriminate when it comes to the types of donated materials he'll take. He says he's often not sure how he'll use the offered materials, but he's open to almost every opportunity: "My friend said, 'I've got a bushel of eyeglass lenses, do you want them? What are you going to do with those?' Well, I don't know, but when you need hundreds of eyeglass lenses, you don't just go out and buy them. The university gave me 15,000 DVDs. Someone offered me a 55-gallon drum of rubber bands once a week. What do you do with that? There's no way to know. But when these things pop up, you say, 'Oh, that's fun, let's try that. These things arrive, and it's as varied as you can possibly imagine."

Dan says the key to creative reuse is the ability to make patterns. Once you have a large quantity of any item, you

One of Dan's design tips is to go for repetition; once you have repetition, you can create a pattern, and once you have a pattern, you have a design.

can make a pattern. Patterns lead to design. He feels that, when put to the task, nearly anyone can do this: "If you have multiples of anything, you have the possibility of repetition. Repetition creates pattern and also unity. Put anyone in a room with a pile of similar objects and say, 'I want a pattern by 3 p.m. or no dinner.' Anyone would come up with a design. It is easy, fun and available to anybody. Most people just don't have the nerve."

Financing the Home

When building homes for very low-income people, financing is always a central issue. Dan Phillips has worked hard to figure out ways to get people into homes without the initial aid of a conventional mortgage. Using the same ingenuity he uses to secure building materials from a variety of sources, he has found a number of ways to finance his projects. One, a partnership with a local outreach program for women in transition, called Brigid's Place, began offering seed money to women building their own homes. Dan has developed a similar program with Living Paradigm, a non-profit in Houston. Living Paradigm maintains the Phoenix Fund, a seed money program that gives future homeowners the money to buy supplies and get started on their home's construction. Because the homes are built so inexpensively and with almost entirely free materials, their value as finished homes is much more than the amount of money they cost to build. Dan says that, by the time the home is built, the homeowner already owns about 80 percent of it outright. Then they seek a conventional mortgage for the remaining 20 percent. He accepts future homeowners with no credit or good credit, but not bad credit.

Dan has worked out alternatives to conventional financing, because, for homeowners with no credit, securing conventional financing is tricky. "If you walk in to meet with a banker and say, 'I've got a great idea to build a house out of trash for people who haven't shown any responsibility,' they're going to say, 'Security, escort this man out,'" Dan explains. "Early out, I had to provide my own collateral," he adds. He's also obtained seed money for projects from private donors. "We've gone to private lenders. A lot of folks can come up with $20,000 and that's all it takes to build a small house. I've had people step up and do that for someone." He envisions communities coming together to create $50,000 rollover funds that pay to start projects, then get paid back by a small mortgage, then fund the completion of more new projects.

Of course, obtaining a plot of land is a key factor for any project. The price of land is one of the major expenses for a potential homeowner, and one of few that is not very negotiable. "Property is always an issue," Dan admits. "You have to buy the property. Once in a while, someone has a lot, but not normally. Generally, you buy property at market rates where you live then what keeps the cost down in the world of Dan is a low-cost labor force and mostly free materials."

Conceiving of how to use all these crazy materials is one of Dan's gifts, according to former intern Jerrod Sterrett, who was nearly finished his builder's degree from SHSC when he interned with Phoenix Commotion. He says the experience taught him aspects of building he didn't learn in the classroom: "I learned more about art than I did about building houses. Dan is a genius in ways people don't really even realize. If you gave most people a bunch of stuff and said, 'Build a floor with it,' they might be able to do it. But he can do things with stuff that gives it this very artistic feel, brings it all together and makes it look like it's not a house made of stuff nobody wanted."

And though he admits his homes are unique in part because of the artistry he and his crew put into them, Dan also credits much of his homes' beauty to the quality of the reclaimed materials he receives: "A lot of these materials

The round windows in the Treehouse's front wall are made of old relish serving plates.

are things even the beautiful people can't afford. They're granite, travertine, vitrified china — wonderful, wonderful things!"

The fun of using unique materials aside, Phoenix Commotion homes also must go through the rigmarole inherent in any building project: efficiency, safety and building codes. Dan is serious when it comes to his homes' efficiency. Low-cost maintenance is a crucial component of making these homes affordable for the Phoenix Commotion's low-income demographic. High levels of insulation, efficient fixtures and appliances and built-in conservation methods make these reused homes run like brand new ones. Some materials weren't meant to be salvaged. "I don't hesitate to buy new," Dan says. "Most of the time I don't have to, but some things I buy new as a matter of policy. I buy new wire because you can't depend on salvaged wire. I buy new pipes for the plumbing, new nails, new screws. Typically I buy new toilets because the salvaged ones I get are the 3-gallon flushers and we're in the world of 1.6-gallon."

A stickler for building codes, Dan states, "You have to build to code. You must pass muster and pass inspection." He bristles at the idea that his homes, or any, should be immune to safety regulations. "Building codes are a good

Dan encourages his crew and the homeowners to create unique designs with the reclaimed materials they've collected.

thing. People who throw rocks at inspectors are being naïve. It's a lot like police officers; we want them around unless they stop us for a ticket. It's the same with inspectors." Meeting code doesn't tie one's hands when it comes to building materials and techniques, he explains, "Every building code has as a provision that alternative materials and strategies are allowed, provided you fulfill the intent of the code."

Dan's unique homes have garnered interest far outside Texas. "We Americans may have invented excess, but the problem of waste is worldwide," Dan says. "I hear from people all over the world asking how to start something like this. I've been featured in magazines in Italy, in Tokyo. I think, Lordy mercy, this is a model whose time has come. I'm not starting anything new. People have been doing this for years — build with what you have."

■ The Phoenix Commotion Builders' School

Part of the way Phoenix Commotion homes stay cheap is through the extremely inexpensive labor performed by a mainly untrained staff. The Phoenix Commotion employs from two to seven workers who work under Dan's skillful tutelage. Dan explains how Phoenix Commotion operates democrati-

Wood log "disks" laid in concrete create a gorgeous countertop.

Phoenix Commotion

cally: "Everyone does everything — from junk work to design responsibilities. No one is immune from doing the junky stuff, including myself." The low labor costs enable him to target his desired demographic of low-income, often single-parent, families and working artists. And hiring untrained builders enables Dan to further benefit his community by providing on-the-job training for in-need Huntsville residents. Though he employs only a small number of employees at a time, the low pay ensures his apprentices are motivated to move on to higher-paying jobs after they've learned their craft. This means Dan leads a rotating crew of untrained workers who leave the program with a valuable skill set and who are instilled with the Phoenix Commotion message of creative reuse and unconventional building methods.

Homeowner and Phoenix Commotion administrative director Kristie Stevens remodeled her own home with Dan in 2008 — a house built previously for another single mother. She says Dan's decade as a dance professor at SHSC is evident in his ease in teaching unfamiliar skills. "Learning how to use all the tools was really empowering," Kristie says. "The way Dan teaches you to do something is by teaching you the most difficult thing first. When he taught me to cut tile on a tile saw, first he taught me to do a circle, which is kind of complicated. You have to make all these small cuts. Once you've done a circle, it's really easy to do a straight line. He'll teach you the hardest part first, so the rest is easy."

Phoenix Commotion doesn't have a tough time finding low-income trainees. Often, they seek Dan out; the program seems to attract them, he says: "They just wander up on the job site. Right now I have a great labor crew. I mean, my, oh, my, they're bright, quick studies. I attract artists a lot." He also doesn't hesitate to reach out to those in need: "I'll see someone who looks down on their luck and say, 'Hey buddy, looks like you need a job. Do you have any issues with drugs, alcohol or the law? No? Then let me teach you.'" His requirements are few, but vital: "You can't have an attitude, you have to have a work ethic, and you've got to be interested in learning."

Dan seeks to instill confidence in his trainees, both volunteers and homeowners, a goal he accomplishes with plain success, as is evident in talking to any of his current or past employees. "Anyone can build. It's all simple," says Phoenix Commotion employee and former volunteer Matt Gifford. He became a Phoenix Commotion intern after hearing about the program while attending the University of Kansas in Lawrence. He says he was fascinated, and headed down to Texas to learn more. Matt quickly adopted Dan's can-do attitude. "People always come in and see a mosaic or something and say, 'Hey, would you come to my house and do this?' And we say, 'No, but better yet,

Dan says people send him wine corks from all over the world, with notes inquiring, "Do you need me to drink more?"

we'll show you how,'" he says. "Anyone can glue mosaic. It doesn't take a special skill set to glue stuff. It's all so simple once you understand the basic concepts of a house and how it works." He suggests that, even if you hire a professional for certain more-intricate parts of a project, you can still do the majority of the building work yourself. "You can't just jump in and do all the more complex stuff, like all the wiring, for example. For that stuff, it's good to have a licensed person, but anything else — floor covering, ceiling covering, wall raising — if you have a back, you can make it work."

Dan expresses his message of self-sufficiency succinctly: "I'm not throwing rocks at builders, but there's not a whole lot to know. Gravity always pulls down. Water runs downhill. If you understand a few basic concepts, you can build a house." For him, the training process is straightforward. "If I say, 'Do this with the saw blade. Do it 35 times, and I'll be back at 3:00.' When I come back at 3:00, you've got that skill. It's that simple." Humility also has a role in Dan's training process. He points out that he was not born with inherent abilities in building—and if he can do it, anyone can: "Nobody is born knowing how to build a house, but you can learn. Anybody can spit on his hands and figure it out."

Though he doesn't think building skills are hard to learn, Dan recognizes that building a home isn't everyone's idea of a walk in the park. Building a house is hard work. But that hard work has a huge payoff. "I'm not trying to oversimplify the process," he says. "It's complicated, messy, dirty, discouraging. But you can live through all that, and look what you get! You get a house."

Phoenix Commotion homes are about as personalized as they come; the Budweiser House is modeled after the classic beer company's designs.

■ Improving Lives: The Phoenix Commotion Homeowner

Every future Phoenix Commotion homeowner is required to participate in the building of her own home. Dan insists upon this practice for several reasons: It intimately connects the homeowner with his dwelling; and it guarantees that the home is customized to its end user from the ground up. He also thinks that when someone works hard for something, it increases its perceived value. Though he believes everyone ought to have the right to work toward a high-quality home, he doesn't think it should be handed to them: "I don't believe people ought to be just given a house. I think they ought to earn it. That's what human beings have done for hundreds of thousands of years." Plus, by participating in building the home, the homeowner knows everything from how her insulation was installed to who delivered the 10,000 frame samples that make up the ceiling.

Kristie Stevens says that building a Phoenix Commotion home improved her life in fundamental ways: "I have two children, and I'm a single parent. It's been a huge deal to have an affordable house payment. My kids have their own rooms and their own bathroom. I have a cute little porch." She was able

Learning to Live Simply

The Phoenix Commotion isn't building luxury homes. The functional, efficient spaces are designed for sustainability in size as much as in materials and efficiency. Downsizing is a growing housing trend nationwide, with the average American home size down in 2010 for the first time in 50 years, according to research by Trulia, a real estate search engine. Smaller homes are a growing trend in part because they are perceived as a route to a simpler life. Phoenix Commotion administrative manager and homeowner Kristie Stevens rented another of Dan's small homes—called the Story-book House—before buying her own Phoenix Commotion home. She recalls, "Before I lived in the Storybook House, I lived in a 1,400-square-foot house that had two bedrooms and one bath, a huge living room and a huge dining room. And I spent the majority of my free time cleaning it. I had two kids and a white linoleum floor, which I despised. When I got ready to move into Storybook House, I had to shed a lot of furniture, some I had paid a lot of money for, some I had inherited. It was painful. It was psychologically difficult to get rid of all that stuff." But Kristie says the payoff was having a more manageable home that allows her to spend more time with her family and less time cleaning. For her, the stuff she got rid of is much less valuable than the time and simplicity she received in return.

to design her home to suit her life: A busy working mother and student, she wanted something small and manageable, but she also wanted her children to have space they could call their own.

Getting involved from the ground up makes for a strong connection between home and resident. It allows homeowners to know their home on a more intimate level than if they move in after it's already built. "You know every nook and cranny," Matt Gifford explains. "Instead of being in this foreign space you inhabit, where you move into a place and only make it your own by putting your stuff on the wall, it's really yours. And you really understand it and you really appreciate it. It's a very intimate thing.

Dan encourages homeowners not only to build their homes, but to personalize them, to connect with them artistically and to use their creativity. Kristie made a giant mirror mosaic in her bathroom. She describes it with enthusiasm: "I did this big half-circle out of mirror, and then there are rays coming out of it. If you walk into my bathroom, above my sink, it looks like this big sunshine with eyes. It's four or five feet across. It pretty much takes up

The bathtub in the Budweiser House is made with a mosaic of scrap tiles designed to look like a frothy mug of beer. The faucet is a reclaimed Bud tap.

Phoenix Commotion

the whole entire wall, but I still want to add some of the same tiles that are on my floor and work them into the sun. It's one of my many projects. When you own your house, you're never totally done. Especially when you know how to do all these different things!"

Intimately designed by and for its unique residents, Phoenix Commotion hand-built homes harken to a time when our dwellings were intimately connected with and reflective of our lives. These distinctly personal homes are not designed for resale 20 years down the road. They are designed as lifelong dwellings. Knowing every construction detail comes in handy if homeowners want to add on to or renovate as family circumstances change. "That's how housing used to work," Matt says. "You'd get married and you'd build a one- or two-room cabin. Then if your wife gets pregnant, you have nine months to build a room for the baby. It reflects your life, and it's a story of you instead of just being a suburban dwelling you get dropped into."

Having hands-on knowledge of one's own home comes in handy whether a renovation is in the works or not. Kristie is empowered by understanding her dwelling, plus she says she saves money on repairs: "If something goes wrong with my house, I try to figure it out or call a friend instead of calling a maintenance guy. It's really great. Even if I can't fix it myself, I will at least know what's wrong. When you live in an apartment, they don't let you touch these things. If you live in a house and don't know what you're doing, it can be dangerous. But I do know what I'm doing. If my faucet is loose, I tighten it up."

Matt Gifford sees the Phoenix Commotion's work with homeowners as part of a bigger change of mindset: Building and maintaining a home isn't beyond their reach. He sees building one's own home as a great first step toward living a self-sufficient lifestyle in a corporate world: "It's not rocket science. That's what we try to teach people. People don't think about picking things up and playing with them like this. They don't think it's possible to do it themselves. They think you need a specialist, because of the way the market is constructed. To break down that divide between 'us' and 'them' culturally, we need to interact. We need community. Instead of our culture being so devoid of self-reliance, we should not be afraid to get our hands dirty, to have our input."

■ Enriching Community:
The Phoenix Commotion Neighborhood

After more than a decade, Phoenix Commotion has had a big impact on Huntsville, Texas. Entire neighborhoods are filled with Phoenix Commotion homes, and their unique style makes them attractive to buyers of all income levels if the original owners move out. Regular people all over Huntsville are eager to move into a Phoenix Commotion home. But Dan Phillips is sticking to his mission of creating homes for those in need. "Homeownership was never meant to be a profit center," Kristie says. "Homeownership should be a right, not something for the elite."

Kristie has seen through personal experience what many studies have shown throughout the years: Owning a home makes people more active members of their communities. "When you own a house, you have neighbors, and your kids play in the street. You pay taxes, and you care that the police are doing their business. It makes you much more invested," she says. Before she built her own Phoenix Commotion home, she rented another of Dan's houses, and says it was the first time she had experienced a secure home and neighborhood: "It was the first time I felt safe letting my kids just ride their bikes down the street. When you live in an apartment, you have to be careful. You don't know the neighbors. Here, we were all neighbors, and we all knew each other. That neighborhood was transformed over the years."

Dan Phillips (back center) and his crew of workers and volunteers.

Living in an established Phoenix Commotion neighborhood showed Kristie a vision of how these homes could transform whole sections of the city. It inspired her not only to want to build a Phoenix Commotion home, but to want to build more Phoenix Commotion neighborhoods. She built her home in a rough area of town, hoping she could set the cornerstone for a neighborhood revitalization. "I knew before I moved in that there were some issues in this neighborhood, but I've talked to the city manager quite a few times because I am a homeowner and I do care. In the two years I've lived here, it's improved dramatically," she says. She fancies taking over the town. "Just two blocks away, one of Dan's former employees is building a house, and there's another employee planning to move into this neighborhood. We're kind of taking over in the same way it happened in that other neighborhood. We're essentially revitalizing this part of the city by putting more homeowners in it."

Along with building safer, stronger communities, increasing the number of homeowners is great for a city's economy. More homeowners means more paid taxes. Providing low-debt homes means community members have more money to reinvest into the local economy. "In Huntsville, there's not a lot of affordable land. Only about 40 percent of people own their homes. It's one of the poorest counties in the state, because not enough people own houses and pay taxes to support the infrastructure to have a sustainable community," Kristie explains. Although sustainable houses are important to a sustainable community, that's not enough, according to her: "A sustainable community

What Did He Do with All That Stuff?

Dan Phillips is a master of creative reuse. Here's a small example of some of the things he's done with unusual items:

- Picture-frame corners from frame stores samples: Colorful, zig-zag ceiling
- Crystal relish platters: Porthole-like windows
- Wine corks: Laid and grouted into floors and countertops
- Wine bottle bases: Stained-glass-like circular windows
- Broken tile and mirror shards: Laid and grouted into floors and countertops
- Thousands of bottle caps: Grouted tile flooring, wall mural
- Broken ceramic mugs: Wall mosaic
- License plates: Roof shingles

- Old cattle bones: Decorative accents, drawer handles, house address numbers, grouted countertop
- Bois d'arc slices and branches: Grouted countertops, stair railings, banisters
- Bondo-filled egg shells: Decorative elements
- Beer bottle labels: Flooring collage
- CDs: Wall and ceiling mosaic
- Papier mâché, sealed: Flooring
- Reclaimed windows: Greenhouse walls
- Champagne corks: Flooring and drawer pulls
- Corrugated metal: Rooftops
- 2-by-4 trimmings: Walls, ceilings, roofs
- Crushed aluminum cans: Shed siding
- Beer keg tap: Faucet

is something close to the amenities you need. It's safe. The creation of more sustainable communities isn't just about the houses, it's about the people."

■ Planting the Seed: Volunteers and Interns

Dan Phillips has trained hundreds of volunteers and interns in the 13 years since Phoenix Commotion's inception. Part builder, part teacher, part social activist, Dan sees the value in spreading his message of reuse, even if it means his building projects require more complex coordination. "From our university in Huntsville, we get students who want service projects, community service hours or just want to volunteer. They come out in droves, which takes a lot of planning, but what you get in return is you have inoculated these people to demand something other than the mass-marketed materials and designs that produce cookie-cutter houses."

In order to make the most of volunteers, Dan has to come up with projects that dozens of totally unskilled workers can learn and repeat for several hours. But his penchant for instruction and the simplicity of many of the tasks required to build a home make the volunteerism pay off for both student and project. "When 25 people arrive on your job site, you better have something for them to do. We often have these people for three to four hours, and there's not time to train them to use a chop saw or nail gun," he says. "But we can organize lumber, build a fence, put down mosaic — there are a number of things that can be done by people who just arrive."

But Dan doesn't take on hundreds of volunteers because he needs someone to build fences and lay mosaic. He knows the most important part of making grassroots change is getting more individuals involved. Every community member who becomes involved with Phoenix Commotion is another advocate for alternative building — or at least someone who has had the opportunity to see a different way of building and creating community. He says, "The idea gets spread, and who knows where that will lead. You get the idea out there. You talk to somebody, and they talk to two people.... it's exponential. It becomes a little groundswell. I'm not saving the world here, but that's the idea."

Building homes for ourselves and others is a direct way to build community. Most of us can say from personal experience that we learn to appreciate and understand one another better after we've worked toward a common goal. Dan says that desire for community is what attracts people to his program: "When we build a house, the volunteers come out. Buried in the most primal parts of us is the instinct to help each other — barn raising, helping someone change a tire. If you can tune into that part of the human psyche, you've done something."

Dan's protegés take with them not just the desire to do good, but the desire to spread the ideas instilled in them through their work with the Phoenix Commotion. For Matt, the biggest impact of the program is "the ability for it to change minds and touch people." He loves the process and the practical applications, but spreading the message is an important part of the overall goal: "You're getting your artistic itchies out, and each house is unique, and that's great. And you're building for low-income families, and that's great. But it's more about all the people who come through on a daily basis and see these ideas."

After Jerrod, currently a professional musician, completed his internship, the Phoenix Commotion way of thinking leaked into all the parts of his life. "The thing I took away was you don't have to spend big gobs of money. Do it yourself. The more you can do it yourself, the more you know about it, and the less it costs you. That's what I think about my band, and that's what I think about houses."

Jerrod attributes Dan's success at spreading his message to his intense passion. "The main thing that makes him so good is that it's what he wants to do. It's his passion. About anybody who takes hold of what they want to do and does it for eight hours a day, I guess they could find a way to make a living out of it if they wanted to. He wanted to build art. He wanted to help people."

CHAPTER 8

MAKING DECONSTRUCTION THE STANDARD MODEL

Creating a national model of deconstructing buildings and reusing their parts offers environmental, social and community benefits to our nation.

Deconstruction — disassembling structures and reusing the building materials — is an idea that has gained popularity over the past 20 years. Still, the number of homes demolished every year greatly outweighs the number of homes that are deconstructed. In conventional demolition, entire structures are bulldozed, resulting in a pile of mixed debris, all of which is sent to the landfill. Occasionally, demolition or salvage crews cherry-pick, removing the most accessible, valuable, easily reused materials from a home, but deconstructing and reusing entire homes is a rarity. Demolition is the standard method of removing structures from building sites.

While it's not mainstream today, "deconstruction" is a relatively new term used to describe an old process — the selective dismantlement or removal of materials from buildings instead of demolition, according to the National Trust for Historic Preservation. Bob Falk, president of the Building Materials Reuse Association, a Chicago-headquartered national non-profit dedicated to raising the awareness and popularity of deconstruction, says it's nothing new: "Decades ago, taking apart buildings and reusing the parts was standard practice. We're seeing a renewal of interest in an old way of doing things."

■ A Rise in Deconstruction

Formed in Canada in 1994, the Building Materials Reuse Association (BMRA) became a US-based organization in 2004. The group's mission is to educate people about the many benefits of deconstruction, and it hosts a biennial conference called DECON where those involved and interested in building materials reuse learn how to efficiently salvage and reuse building materials and how to spread deconstruction's benefits. The BMRA isn't alone. Groups fostering the reuse of materials and businesses that sell salvaged building materials have been springing up with increasing regularity all over the US. Habitat for Humanity launched its first ReStore just 20 years ago; they sell donated and salvaged building materials from deconstructed and remodeled homes. There are now more than 700 ReStore locations. The BMRA estimates that hundreds of other for-profit and not-for-profit businesses and organizations countrywide revolve around the use of reclaimed building materials.

Our nation experienced a change in mindset after the Industrial Revolution. Before the advent of mass production, we valued craftsmanship and high-quality materials because items were expected to last a lifetime. But cheaper materials and an increase in our speed of production shifted our values toward efficiency. It's led us to develop practices that favor expediency over high quality. "We used to put a higher value on craftmanship and quality materials in contrast to today's focus on profit-based efficiency," Falk says. "Also, there has been a change over the past several decades, especially since World War II with the development of heavy machinery and worker safety laws." Though worker safety is paramount, safety requirements have resulted in rules that keep workers in machines, rather than allowing them to use their hands and disassemble and recover usable parts. "It created a disincentive for salvage and reuse," Falk explains. "The tendency has been toward tearing something down with a machine and throwing the materials away rather than taking it apart and reusing the materials." The BMRA was formed to help reverse that trend.

■ Environmental, Community and Economic Benefits

The BMRA advocates the many benefits of building materials reuse. The first major benefit is environmental. Building materials reuse is a means of "con-

Guy and Kay Baker estimate they spent $20 on the kitchen; hinges and doorknobs were the only things they paid for. Kay laid the wood-block countertops herself.

The Bakers love spending time outdoors, and designed their home for comfortable, year-round outdoor living.

Photos: Michael Shopenn

The Baker home is something of a museum of its region's architectural history. Guy says the home includes something from every town in his county.

Guy had a cattle trough lined with fiberglass for a huge tub with a $90 price tag.

Photos: Michael Shopenn

Meghan and Aaron Powers built their Idaho straw bale home with the help of friends and family who camped out during construction.

A grain bin was converted into a combination office/workshop and guest house.

The Powers employed a number of clever space-saving techniques in their 836-square-foot home, like the sunken dinner table beneath removable planks of the living room floor.

Photos: Betsy Morrison

Rick and Naomi Maddux' handbuilt, straw-bale home was a work of art they built with friends and neighbors over several years, and with no construction loan.

They wanted their home to be unique and imbued with character. One of their design requirements was to avoid hallways.

Photos: Paul Weinrauch, courtesy *Boulder County Home & Garden* magazine.

Inside the Phoenix Commotion Treehouse Studio, frame corner samples from a photography framing store combine to create a zigzag-pattern ceiling.

Photo: Courtesy Phoenix Commotion

The Storybook House is one of dozens of Phoenix Commotion homes that are taking over whole sections of the city.

Wood log "disks" laid in concrete create a gorgeous countertop.

Photo: Courtesy Phoenix Commotion

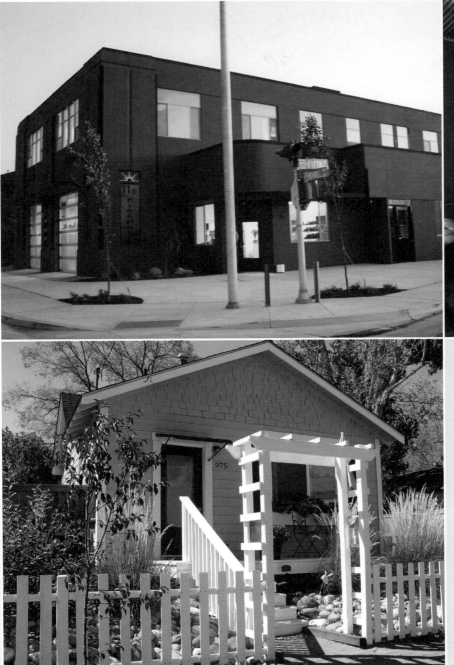

HabeRae rescued an abandoned 1950s firehouse in a dangerous part of the city. Now a 5,000-square-foot mixed-use building, the space offers a bagel shop, a salon and nine residential lofts.

A classic example of urban infill, Haberae's "2 on Watt" project was built on an abandoned lot in an old part of Reno. The cottage style was designed to fit in with the homes' historic surroundings.

Photos: Pamela Haberman

Builders of Hope rescues scheduled-for-demolition homes and either renovates them in place or moves them to whole new neighborhoods made up of buildings they have rehabilitated.

Design features such as front porches and positioning on cul-de-sacs increases the sense of safety and community in Builders of Hope neighborhoods.

Photos: Courtesy Builders of Hope

serving our natural resources by preserving quality materials," Falk says. "Most of the lumber crunched up in home demolition is our old-growth forest, and that will never be available again." By reducing waste and reusing quality materials, we take advantage of the resources of the past and reduce the need for new materials production.

Building matrials reuse also benefits communities in other ways. Building a thriving deconstruction industry would create jobs. Deconstruction and the process of transporting, cataloguing and selling used building materials require more human involvement, and therefore more jobs, than demolition. "Rather than one guy in a big excavator that's crunching a building, and another in a truck that's hauling it to some distant landfill, you can put the cost of that building removal, as well as the value of the salvaged materials back into the local community in the form of more jobs," Falk states. The tasks involved in the emerging deconstruction industry require a wide range of skill and interest levels, providing jobs to members of all levels of the labor force. "One of the advantages of deconstruction is that it's not rocket science," he adds. "Taking apart a building and salvaging materials is less complicated than constructing a new building. You can offer people that have few job skills a means to learn some basic job-site safety, working with and around machinery, proper tool usage, and the like. So, the actual deconstruction process itself can help train people for jobs in the growing deconstruction industry or in the broader construction industry." Builders of Hope offers a great model of an institutionalized system of deconstruction and community on-the-job training (see Chapter 9).

Along with jobs that deal with the hands-on process of deconstruction, reuse also supports ancillary businesses that process and sell reclaimed materials. "You can sell the materials you salvage, so there's an opportunity for people to start small businesses, which teaches how to inventory and manage materials, and how to distribute and market them," Falk says. Businesses already working in the industry range in size and scope, and there's plenty of room for increases in these types of businesses. "This is not necessarily just a little corner store that sells a few hundred dollars' worth of materials," Falk explains. "Some ReStores, dealing with both donated and deconstructed materials, do $2 million worth of business a year. So we're not just talking about low-end, unsustainable businesses. Interestingly, and in spite of the down economy, many reuse businesses in our membership have been reporting very positive sales. I think more and more shoppers are realizing that quality materials can be found at reused building materials stores and at lower prices than the big box stores."

Salvaged building materials are valuable, and urban centers can directly benefit from keeping that flow of resources in the community. "It's an economic benefit to the local community, rather than all of that value being exported to the landfill and being forever lost," Falk says. In many cases, the urban centers with the most abandoned buildings and with the greatest need for deconstruction services are also those with the greatest need of low-skilled employment opportunities. "Many of our eastern Rust Belt cities, where there has been mass migration from the inner city and there is a lot of abandoned housing that needs to be removed, are the very same places that have an employment problem. With an emphasis on employing people, deconstruction can be an important element to help transform distressed neighborhoods into more sustainable communities," Falk suggests.

■ What Deconstruction Builds

When resources are kept within the community and become the catalyst for new industry, the entire city benefits. "When you have a business that's employing people, it provides a larger tax base for the community. It employs people, ups their standard of living and provides materials for the community to rebuild infrastructure," Falk says. He sees deconstruction working hand in hand with other community development projects to create communities that are more self-sufficient: "In the inner cities, we're seeing this growth of urban agriculture and creative reuse of the abandoned land. Deconstruction dovetails with these other efforts to create opportunity in areas where there hasn't been much hope."

By reusing building materials within the community and hiring local community members to both dismantle and rebuild, communities can create a closed loop of reuse within communities. "It provides materials for construction, in the same way these urban farms provide food for their communities,"

Falk explains. "It's one element in developing a sense of community, rather than being totally dependent on outside sources for food and building materials. There's a richness of materials in these buildings. Some people just see it as trash, but there's a lot of intrinsic value. I think all of us feel better about using things that we produce or generate locally than about using something that comes from a far-away factory." Communities also benefit by retaining elements of their shared past, and holding on to longstanding cultural elements. "There's a way of preserving some of the community history by reincorporating these materials back into renovation in that community," Falk adds.

Implementing deconstruction as an alternative to demolition has been a challenge in many cities. "Deconstruction is still foreign to a lot of people who regulate building removal, however that trend is changing," Falk says. While more cities are becoming deconstruction friendly, the licensing requirements, training requirements, and building removal ordinances in most need to evolve to better accommodate deconstruction and building materials salvage as an alternative to demolition.

But Falk is optimistic. Awareness of deconstruction as a viable method of building removal is growing. "While not mainstream nationwide, more and more people are looking at deconstruction as a building removal option to demolition. Our association has gotten more attention and our membership is growing steadily," Falk reports. As cities take on their problems with blighted inner-city areas, deconstruction is an idea that's spreading fast. "We're getting a lot of interest from municipal folks who deal with redevelopment efforts and workforce development. They are starting to understand the connection that deconstruction can make between building removal efforts, job training, small business development, and community redevelopment. Some cities are pretty progressive, such as Seattle or Portland. However, interest is growing on the East Coast as well. Middle America is catching up and there are centers of real interest. In Iowa, for example, we're working with the state, and they've been very open-minded and forward-thinking. We're making good progress, but like any effort at social change it takes some time to get the habitual ways of doing things to change," he says.

CHAPTER 9
BUILDING HOPE

A North Carolina program rescues landfill-destined
homes and gives them new life as foundations
of low-income neighborhoods.

I N NORTH CAROLINA, the Builders of Hope program rehabilitates rundown homes, creating eco-friendly dwellings for low-income families. In 2006, after inheriting a small amount of investment capital, Nancy Murray left a career as an advertising executive to found the non-profit. "I was at a point in my life where I had the opportunity to do a career change," Murray says. "I always had the inspiration, even as a child, to make a difference in the world, but I didn't know how. My father passed away and left me some money, and I thought, 'Well, I could put it in the stock market' — not knowing it would crash — 'Or I could save up for retirement, or I could do something significant.' My husband and I decided I should start a non-profit to help others. We were in the position to do that, and I felt very fortunate."

Murray wasn't exactly sure what type of non-profit to start but says, as she was considering the question, it seemed as if the universe was working to demonstrate to her the severe lack of low-income housing in Raleigh. She continually met people in need of high-quality housing near the city center where they worked. "I started it as a ministry," she recalls. "People kept coming into my life, either through church or other groups, who had to travel a long way every day to get to work." She started looking into the local housing market and considering how difficult it was for working, low-income people to make a good life for their families in the city, given the available housing options. "Really, for the last 30 years, if you're a teacher, firefighter, secretary, worker at a non-profit — any careers that are really great careers but not on the higher end of the pay spectrum — you have the choice: It's either substandard housing in the city; or you can live in the suburbs, get some property and one small vinyl house and commute."

Once Murray started to realize how few housing options existed for low-income urban workers, she became interested in how much housing was going to waste in Raleigh. She started noticing new buildings being built around the city and began to pay attention to what was happening to the existing homes. "Things started coming together," Murray explains. "All the rich neighbor-hoods were tearing down houses left and right with wood floors and granite counters. The land was worth more than the house, so the house would go to

Builders of Hope founder Nancy Murray realized the connection between the number of homes being torn down and the number of people in need of affordable urban houses, and she designed a non-profit that brought the two together.

Builders of Hope

the landfill. Then I realized how many: 40 to 50 houses a month just in the one town where I started!"

Murray quickly envisioned an elegant solution that helped solve two problems at once: All the homes being sent to the landfill were the perfect fodder to create homes for all the people in need of decent places to live. And with that, Builders of Hope was born. Murray felt dedicated to three goals: first, to provide homes for working families; second, to keep all those high-quality, usable building materials out of the landfill; and finally, to make a difference in the greater community by providing jobs for unskilled workers. "I realized, first off, our planet cannot sustain this," Murray says. "There are more people without housing than ever before, and we're sending all these houses to the landfill. And I wanted to make a difference not only for the people who moved in, but also those who build them, so I decided to hire people from homeless shelters."

■ A Non-profit Is Born

In 2007, Murray founded Builders of Hope, a non-profit that purchases teardown homes, rehabilitates them and sells them as low-income housing. Committed to sustainability from the recycling of the home to the health of its materials, she creates durable, energy-efficient homes, working with national organizations and programs such as the US Green Building Council and LEED. "The Builders of Hope model has successfully been able to address the major issues: How do you take this inventory and make it beautiful and sustainable and eco-friendly," Murray reports. "We work with LEED, so we're energy-efficient. We strive for superior indoor air quality. Everything is about the end user — the tenant or the homebuyer."

It has absolutely worked, confirms homeowner Josh Thompson, who moved into Builders of Hope's State Street Village in Raleigh in May 2010. He describes living in a Builders of Hope home: "It's a phenomenal experience. The overall quality of the home has made me feel confident that it's going to endure." He adds that the quality of materials in this home compared with other affordable places he has lived puts his Builders of Hope home in its own league: "There's so much cheap construction out there, homes built of concrete slab or vinyl siding, but Builders of Hope uses superior-quality materials. It makes the experience awesome."

Builders of Hope homes are special to homeowners for reasons outside their longevity and high-quality materials. They love their homes because they are custom-built to fit the end users' specific needs. "The design of our house is not excessive," Thompson says. "But they took the time to figure out my family and priorities in life, and they designed a house around that. They say, 'Oh, you want kids? Let's give you big bedrooms. You want to host events? Let's give you lots of space.'" Thompson and his wife host weekly church get-togethers for 30 to 40 of their friends, and he says it's quite comfortable in their 1,600-square-foot home, thanks to the Builders of Hope designers' attention to detail and wise design.

Murray carefully considers the elements that go into her homes, examining them in terms of durability, efficiency and health. Making homes work for homeowners is her end goal, and that sometimes means choosing among many sustainable building options. She designs every home with the hope that one family will live in it for 20 or 30 years or more, meaning she often chooses quality over low cost, or durability over environmentalism.

Murray learned that we demolish 250,000 homes a year, and that we send nearly all of those building supplies to the landfill.

One example is the selection of exteriors. No fan of inexpensive vinyl siding, which she says only lasts for about 15 years, Murray uses only Hardiplank siding on her homes, an eco-friendly, durable option, but a more expensive one. The higher-quality materials pay off over time, Murray says. By choosing Hardiplank, she can be assured homeowners won't have to worry about a dilapidated exterior in need of replacement in a few years. Murray cites another example: "All the houses need insulation. There is generally no insulation in homes built prior to 1965. We use blown-in foam insulation. It's more expensive, but it gives you the opportunity to really seal the house. It's superior-rated in terms of insulating quality." That means they will be less expensive for homeowners to maintain over time, something Murray considers equally important as a low initial price.

Murray also insists on quality over quantity for her homes' heating and ventilation systems, a crucial component of healthy indoor air quality: "You must get a high exchange rate on the HVAC [Heating, Ventilating, and Air Conditioning]. You pay more to update the HVAC with a higher indoor/outdoor air exchange, but then if there's any yucky stuff inside, you're not sealing it all in."

And though Murray is always willing to spend more if quality is at stake, she is also willing to forego efficiency-enhancing systems if it will sacrifice longevity or economy. "This is an example," Murray explains. "You get big LEED points to be rated for solar, but solar is ridiculous for us. There's the

Updated with durable, efficient, modern amenities including new roofs, efficient windows and cutting-edge HVAC systems, Murray designs her rescued buildings for the families who will live there.

opportunity for it to get broken or smashed. In some neighborhoods, it's a high likelihood." Though her homeowners could save on electricity by using solar power, the expense of maintenance for complex technological equipment doesn't work for her specific clientele. "Even if they don't get damaged, in 15 years, most solar panels will have to be replaced. I have to ask myself, 'In 15 years, can that homeowner spend that amount to replace it?' Chances are no — it's not a sustainable solution, so we have to come up with a better way."

Though compromise and tough decision-making are always necessary in Murray's world, homes must meet the highest green standards possible, and she continuously quests to perfect the product she provides: "Part of all sustainability is really challenging each other, and not accepting no for an answer. If we haven't figured something out, it just means we haven't figured out how to do it yet. We're constantly challenging the people we work with to keep it green and affordable."

■ Sustainability at Large

Builders of Hope makes big headway on the sustainability front by starting with reclaimed housing stock. And, though she's committed to the health of her future homeowners and their ability to maintain their homes over time, Murray's interest in sustainability extends beyond her homes' residents. She also hopes to benefit society at large by confronting the huge housing waste problem in the US: "We're taking a house that's been rendered functionally obsolete, and we're completely rebuilding it, turning it back into tax-generating revenue stream and offering it up as affordable housing."

Builders of Hope rescues scheduled-for-demolition homes and either renovates them in place or moves them to whole new neighborhoods made up of salvaged buildings.

Murray believes increased home-ownership leads to proud families and healthy communities, and she knows healthy housing for low-income residents is very hard to find. "In every city and even rural towns, there is a huge shortage of affordable housing. Raleigh is 23,000 homes short, yet homes are sent to the landfill every day," she reports. "There are 33 million homes across the country that were built before 1970. That's 40 years or more of housing stock, and there's no national program for rehabilitating it and making it green."

If Murray is amazed at the amount of waste we're sending to the landfill, it's because she's seen how much of it we can save and reuse, adding that "25 of our houses represent 1.5 million pounds of debris rescued from the landfill." Murray says when we compare the amount of homes we toss in the garbage heap with the numbers of new homes we need, a fundamental paradox is revealed: "Of the 33 million pre-1970 homes, 250,000 are torn down every year, and those are just the ones tracked by the EPA. There aren't enough holes in the ground to be able to withstand this. It's a teardown epidemic. It's not sustainable, and as a country we need to figure out a better way."

■ Builders of Hope Homeowners

Murray aims to provide quality homes for the demographic of families that don't qualify for most social programs, but that can't afford a home on the standard market. "All of our buyers hope to get conventional financing," she says. "Our houses range from 60 to 120 percent of the median house value. Habitat for Humanity aims at about 35 to 55 percent of the median house range, and most of their clients are people who can't get mortgages." Murray targets low-income working families who have been pushed away from the urban centers where they work. "They're often first-time homeowners who have grown up on public housing. They're young families who can't afford anything otherwise."

Dawana Stanley is president of the Barrington Village homeowners' association. Every Builders of Hope neighborhood has community groups that help residents get to know one another by hosting parties and study groups and connecting with outside community groups.

Homeowner Josh Thompson says the location and design of his home has changed his family's life: "We have friends in the neighborhood, and there are so many other children. The location has changed my quality of life." Thompson appreciates that his home is located not only in a safe, friendly neighborhood but also in the urban center, where most of his activities are located. He feels that living in the city helps him teach his young children to be thankful: "My kids get to see diversity, because our home isn't isolated away from poverty. They aren't growing up in an environment that says, 'This is what you automatically deserve.' They get to see that they've been given a gift of a quality house, and not everybody has that." A city-lover, Thompson didn't want his kids growing up in the isolated suburbs, where every neighborhood the family could afford was filled with mass-marketed, cookie-cutter homes filled with cheap plywood and vinyl. Living in a Builders of Hope home teaches an appreciation for reuse, and for giving back. "Even at my son's age of three, he understands that Christmas isn't just about getting presents," Thompson says. "He sees that the people across the street don't have something, and he asks, 'What we can do to get them some gifts?' Builders of Hope lets me teach my kids that."

Murray also appreciates the lessons she's learned through her work with Builders of Hope. She says the program has shattered any remnants of naïveté from a sheltered upbringing: "My dad raised me to be white collar, with a big-business mentality. We're an IBM family. Growing up on the other side of the fence, I never really understood the impact that housing has on people. I grew up thinking substandard housing is like that because the people who lived there didn't take care of it. But it's like that because the landlord doesn't take care of it. People live where they have to. You might say they could live elsewhere, but they can't. Every city has shortfalls of affordable housing by thousands of units."

Murray is committed to putting families into high-quality homes they can afford, and she has included several fail-safes to encourage and support families so they are able to stay in their homes for many years. "We sell the houses at our cost. Let's say a house was appraised for $150,000, but our cost to build was $135,000. We would sell the house for $150,000, but we would only require the homeowner to bring $135,000 to closing," she explains. "We take the extra, and we hold onto the second mortgage, which is 100 percent forgivable. Every homeowner has closed with equity on the table." The second mortgage, which covers the difference between the home's construction costs and appraisal value, is forgiven over time as the homeowner stays in the house. After five years, a portion is forgiven. If the homeowner stays in the home for ten years, the entire second mortgage is forgiven, increasing the home's equity.

■ Builders of Hope Neighborhoods

Murray has two models of building Builders of Hope neighborhoods. The first is to buy scheduled-for-demolition homes in the inner city, rehabilitate them where they stand and make them the cornerstones of entire renovated neighborhoods. The second is to pick up several tear-down houses from different areas of the city and move them to create new urban neighborhoods made entirely of Builders of Hope homes. When homes stay in place, Builders

Design features such as front porches and positioning on cul-de-sacs increases the sense of safety and community in Builders of Hope neighborhoods.

of Hope often tries to buy five to ten homes on the same street and build a cohesive community.

Murray wants to improve more than the individual homes her residents can afford in the city; she wants to improve the quality and safety of the communities they can live in, as well. Though she was aware of poor inner-city conditions, she says that until she started working in a field related to substandard housing, she hadn't realized how poor inner-city living conditions could be: "To give you an example, in many of these places, you have to bring your own appliances, including stoves. In one unit, a window air conditioner was stolen and never replaced. There will be gaps under the door big enough that rodents can just walk right in. These are people who are working, not just those without jobs. We find people living in crawl spaces."

Working in blighted neighborhoods and getting to know their residents has shown Murray that the structure of the current urban housing market is the root cause of much of the destitution. Inner-city low- to middle-income people can't afford decent places to live, which leads to community-wide despair, lack of civic involvement and increased crime levels. "Right now, we're revitalizing 40 apartments and 25 single-family homes on one street. On this street, there's prostitution and a market where they are selling drugs," Mur-

Building Green

Builders of Hope lists its key green building ingredients on its website.[1]

KEY GREEN FEATURES

- Recycling existing homes and useable materials; an average of 65 percent of the structure is reused
- Develop infill locations when possible
- Sustainable building practices and material
- Passive solar orientation
- Spray foam insulation
- Exterior ventilation
- Fluorescent lighting
- Low-emissivity (low-E) windows
- Low-flow plumbing fixtures
- Energy Star appliances and water heater
- Low-VOC materials and sealants (VOCs are unhealthy chemicals that outgas at room temperature from building supplies and interior finishes)
- Upgraded air sealing, insulation and HVAC
- Large front porches to reduce heating and cooling bills
- Rain barrels and drought-tolerant landscaping

ray says. "There's a guy who is about 70 years old who comes to one of our sites, and he calls me Mother Teresa. He said, 'You are the first one ever in my entire life that has come into this neighborhood and built something for us. Every time someone comes and builds something new, it's something we can't afford.'"

Murray wants to provide high-quality housing for the people who already live in urban neighborhoods — not to price out current residents and bring in higher-income homeowners. "We're about going into the neighborhood, hiring from the community and building for the actual people. These ghettos were formed with invisible walls and barriers. When all the people moved out and built the suburbs, no one said, 'You can't live here,' but it was priced so they couldn't afford to live there," she says.

Murray designs her renovations with community-building in mind. She always includes front porches and communal spaces and develops neighborhood associations where homeowners get together to discuss safety and other issues. "They work hard to build community, and they do it well," Thompson says. "They invest extra to give everyone the Internet and a neighborhood communication system, which helps facilitate community immediately. And they have cookouts where all the neighbors can meet each other." Builders of Hope also helps residents connect with the wider community. "They worked hard to connect me with leaders in the community," Thompson adds, a task that originates for Builders of Hope long before a homeowner moves in. He explains, "They are really involved in the community. Let's say the community has an activity council. Builders of Hope goes out of their way to attend the meetings long before they build a community, so when they start the community, they can plug you into it through the contacts they've developed." When he moved in, he met area leaders, teachers and pastors. The revitalized communities are, by design, connected neighborhoods where residents know each other and watch out for each other's safety. "This tells you the comfort that people feel in my neighborhood," Thompson says. "My son loves to dress up as an Indian. So here's my son — he's three — in the front yard, shining like a light, wearing nothing but a loincloth. It's an example of how safe my kids — and I — feel."

Murray is hoping to revitalize urban neighborhoods and to buy up as much inner-city property as she can: "Now that it's cool to live downtown again, the land is cheap, and it's getting scooped up." Developers in most cities buy up cheap land, then build high-end housing there, forcing out longtime residents, and she wants to change the status quo: "These people have lived there their whole lives, then someone comes and makes it beautiful and they are booted out. I'm saying, 'We're going to make it beautiful, and guess what? You can stay.'"

■ Building Inspiration

Though she knew building homes would have a positive effect on future homeowners, Murray says she didn't fully realize the far-reaching effect that improved low-income housing has on entire communities. "There's another guy who walks around, he is 70-something, he said, 'I never thought I would be able to walk around this neighborhood and feel safe again. I haven't felt safe for ten years!' You think you are doing it for the people who will live there, but it's the whole community that benefits."

People who grew up in and moved out of city slums are often thrilled at the prospect of returning to their childhood homes revamped. The people attracted to Builders of Hope communities are frequently people who grew up there and were forced out by unsafe, substandard living conditions. "Folks want to live there again because they grew up there. They want to settle down there again," Murray says. "I didn't anticipate how the fact that you're not tearing down, you're rebuilding, is very inspirational to those who have lived there in the past."

Murray also strives to encourage further community and economic development by partnering with companies that normally serve higher-income areas. "Builders of Hope says, 'What can I do to bring in such and such a company that normally wouldn't work in this part of town,'" Thompson explains. "They show them these communities, and the opportunities here. A number of suburban companies have been brought into the city."

The program also connects people from different communities and benefits the people who donate homes. "Through a friend, we met the people who donated our house," Thompson says. "We randomly went to a get-together at this person's house, and the person who donated our house was there. From their perspective, they felt the benefits, too. They saw who they were able to help. They saw how their act of sustainability provided an all-wood-floor home for a family. All around, not just from my end, but from the people donating, it's an awesome experience."

Despite her wide-ranging efforts to help individuals within her community, Murray might find the most satisfaction in simply telling people currently living in extremely poor conditions that they can move into a healthy new home they can afford. "We worked on these substandard units — out of the 40 units, 20 were occupied and 20 weren't. First, we rehabbed the empty ones, then we gave current residents the opportunity to stay for the same price," she says. "Only now, they had a LEED-certified gorgeous home with things like bamboo flooring. You should see the expressions on their faces." Having a pride-inspiring home completely changes homeowners' lives, Murray says, adding, "They've decorated, their kids bring people over. When you give people a home that breeds dignity, they also get hope and inspiration. It's amazing what an impact a beautiful home someone can be proud of has on their life."

■ Builders' Builders

Nancy Murray knew that another problem for many low-income workers is a lack of training. For those who can't afford or don't want to attend colleges and trade schools, it can be difficult to get high-quality work experience. She also knew that hiring a partially low-pay work staff would help keep her homes affordable. She started the Hope Works program, through which she partners with social service providers such as local rescue missions, workforce

Funding Hope

Murray built Barrington Village with no outside financial assistance. She saw that other groups who receive government subsidies to build low-income housing weren't using their resources efficiently, and she knew doing more with less was the most surefire way to demonstrate a better model. "The first neighborhood I did, we used no government funding. I wanted to prove I could build affordable housing without handouts and do it less expensively than those getting subsidies."

Oh, Volunteers!

Cheryl Cotter is the service learning coordinator at Carey Academy in Carey, North Carolina, just outside the Raleigh-Durham metropolitan area. Carey Academy provides grades 6 through 12 education, and Cotter frequently sends large groups of students to volunteer with the Builders of Hope program.

Carey Academy students were among the first volunteer groups to work with Builders of Hope, when it was a fledgling program without a formalized volunteer program. As the non-profit's success and size grew, so did its volunteer program, and the newly formalized volunteer program's coordinator contacted Cotter to see if she'd like to set up a regular volunteer schedule for students. "I'm always looking for new opportunities that demonstrate the full circle of being an active citizen at any age," Cotter says. "That's why I really like Builders of Hope. Here is a woman who started with a dream and took it from there," she says, referring to Murray's can-do, make-it-happen attitude. Builders of Hope created a program called Super Saturdays, through which large groups of volunteers converge to help on large projects. Cotter sends 10th, 11th and 12th graders every year, and sends additional students in smaller batches throughout the year. She says all of her students rave about working with Builders of Hope: "They love the demolition…whacking a wall. They love taking something apart and then seeing down the road what became of that place. I have yet to hear any of our students dislike or have a bad experience with Builders of Hope."

Though Carey Academy has no requirements for community service, the school has worked to develop a culture of volunteerism, and nearly every student participates in volunteer activities. "Here in Wake County, the public schools require volunteerism for graduation. We do not," Cotter says. "Our school isn't big. The high-school portion is about 425 students—about 100 per grade, 9 through 12. Of that, almost 300 are in a service club. They choose to be in those clubs. They don't have to be." Cotter says Builders of Hope's location amidst three college towns is advantageous, because the colleges encourage a regional spirit of giving back, and her students dig right in: "It almost seems like osmosis sometimes. They see it, they hear it, they go do it once or twice, and then all of a sudden, it's who they are and what they do."

Cotter believes part of the reason students become particularly engaged with Builders of Hope, and perhaps a key to the program's success, is its emphasis on appreciating its volunteers, and on showing volunteers the value of their efforts. "Their workers at the site have always been the best at taking time to tell the kids what a great job they're doing," she says. "They always give heartfelt thanks and explain to the kids how important it is that people volunteer. We hear all the time from them, 'Wow, you're doing that a lot faster than I thought you would!' The kids feel they've really contributed and made a difference."

Cotter loves the sense of empowerment her young students get working with Builders of Hope. The program provides a rare opportunity for hands-on learning for students who spend much of their time in the classroom. "They learn that they have a new skill that they didn't know they had," she says. "Just that they learned they could tear out nails or hammer in nails.... That is something they will keep with them when they have their own home or apartment. They are empowered. I think that's why they always want to go back and do more."

Cotter says a few particular students stand out when she thinks about her work with Builders of Hope. "I can think of one very quiet young man who was a sophomore—his inclination was not to get out and do service. He hadn't done it before. When we got there, he was the one out there getting the muddiest, the dirtiest, hefting around more sod than I've ever seen!" she adds, laughing. "Another was a young man who's diabetic. Sometimes I think he had limited himself, but he was hefting this pickaxe to dig a ditch for the water piping. He was just hauling that thing around and having a ball. He was grinning ear-to-ear the whole time."

Carey students do a wide variety of tasks on Builders of Hope sites. "They do landscaping. They do painting. They've done light trim work and framing," Cotter says. "They can't use power tools. It's mostly the smaller interior stuff, like nailing. Their favorite thing is deconstruction." She teaches her students that every task on the job site is important. The program teaches students flexibility, and about the value of hard work. "When you volunteer, the first thing you learn is to be flexible," she says. "When we're there, we're there to be of service, and to do whatever it is they have determined we need to do." Regardless of the tasks assigned them, "they always want to go back. Always, always, always."

Cotter also likes that volunteering gives the students lessons in being a good citizen. She frequently invites guest speakers to her class, and a leader at the North Carolina State Service Department comes to speak to the students every year. "He has always stressed to the kids, don't wait until you have a job and are a taxpayer. You can make a difference now," she recalls. "He says learn about the community, what its needs are and what interests you, and your community will be better for it." She says the sense of empowerment they feel when wielding a hammer is doubled when they consider the difference they can make with their efforts: "You don't have to be a taxpayer to have a voice. Everyone can have a voice, even at a young age." She tries to show her students additional ways to impact their world. "We do letter-writing campaigns. It's not just about service, but also about civic action—find out who your congresspeople are, see how the system works, and see how you want to have a voice."

For Cotter, the best part of working with the Builders of Hope volunteer program, though, is when the students see the end results of their efforts and truly understand the value of what they've done for another family. "Their eyes are opened to the fact that all humans are the same, and when we all come together, we all can benefit from it. When they're out there throwing sod around and getting muddy or inside painting or ripping it apart, they don't totally see it. It's not until the end when it's finished and they meet the family—that's when it really hits them," she says.

development organizations, the Department of Corrections and the Veterans Administration. Program participants learn a high-quality, marketable skill set in construction and remodeling. Because Murray's homes are built to high green standards, graduates are also well-versed in sustainable building and in renovating homes to meet various green certifications. After six months, participants graduate with a certification as a construction framer and a recommendation for hire from Builders of Hope. The social service providers that Builders of Hope works with to obtain students also help graduates find quality employment. Murray appreciates that, along with providing homes and safe neighborhoods, her program provides an avenue to high-quality employment, and helps reduce the number of people in need of government financial assistance.

Murray's commitment to the end users of her homes is evident in her commitment to financial efficiency and high output. She says her program is designed to produce homes, not make tons of money or pay high salaries: "If you want to do a comparison, take a look at Habitat for Humanity. They're in multiple cities, and collectively, they build a ton of houses. But the thing is, they work much like the government. They're very fat. They have an operations budget of about $2.5 million, and they build about 12 to maybe 25 houses a year. On top of their operations budget, they get all their materials for free, and they get every house sponsored. Comparatively speaking, we have an operations budget of $2.9 million, and we're building more than 200 units a year, with no sponsors."

Builders of Hope is able to produce homes so efficiently by performing every task related to the building of the homes, from homeowner applications to design and construction. "We work directly with the homebuyer, and we are the builders. We've got our realtors' license, and we've got the architect, land planner and legal counsel all on staff. We wear a lot of hats. We work smart," Murray explains. Because every aspect of the building process is centralized, Builders of Hope also eliminates the extra costs that often add up as multiple subcontractors work on conventional home sites. "If a homebuyer comes in and wants to make a building decision, we're the builder, so we can tell them right on the spot — 'Yes, we can move this wall,'" Murray says. "Usually, you'd have to send the question to the general contractor, then it goes to the architect, and along the way, everyone adds on 20 percent."

Working efficiently and wisely is only one of many ways Builders of Hope works to keep its homes affordable. The greatest single contributor to creating low-cost homes is starting with donated housing stock. Every Builders of Hope home starts with a donated home that would otherwise have been torn down. Murray says getting homes donated is no difficult feat: "All of our houses are donated. People would normally have to pay $10,000 to $12,000 to have an older home abated and torn down and the site cleared. But if they donate the house, they get a huge tax write-off for the value of the house. Once it's moved onto a new foundation, that home is an asset with no debt." Then Builders of Hope takes out a line of equity to rehabilitate the home, and repays the bank when they sell the home.

After proving Builders of Hope could operate with no outside assistance, Murray decided to see how the various programs designed to help non-profits could benefit her output and productivity. Several nationwide programs help provide building supplies to non-profits. "Home Depot has a program called Framing Hope, and we get free vanities and free toilets through that program," Murray says. "There are other programs, like the [North Carolina] Housing Finance Agency. They pay $4,000 toward sustainability upgrades for a home — there are many statewide programs like that for non-profits."

Murray's success meant that government agencies started coming to her. "After Barrington Village, the mayor was so excited, he asked us to do a model of affordable housing for the City of Raleigh," Murray says. The City had already put together an eight-acre patch of land for the project in the middle of Raleigh. The mayor offered Builders of Hope a zero-percent financed acquisition of the land, which allowed them to begin multiple projects at once. "It's huge for us. It allows us to do multiple projects at once because we don't have to buy the land upfront." Working with city governments allows Builders of Hope to be more prolific, and sometimes to provide homes for less money than they could otherwise. Other counties and cities, including Durham, North Carolina, have offered land deals to Builders of Hope. "In Durham, they have community development block grant funds," Murray says. "We buy land from their land bank, then the city council votes to pay us back for the land. The City pays for the acquisition, then we pay for the rehab, so it allows us to sell all the houses for $80,000 to $90,000."

Murray is determined that Builders of Hope houses be extremely affordable. She says other groups often get subsidies and plan to build affordable housing, but unplanned building costs and higher-than-necessary company profits mean the homes end up costing more than most truly modest and low-income families can afford. "Take East Lake in Atlanta. Everyone held it up as the most amazing neighborhood ever. They built this apartment village, but for a two-bedroom, the cheapest rent is $850 a month. The median income there allows for about $500," Murray says. By reusing homes and building supplies, working with government agencies and procuring a low-cost workforce, Murray creates higher quality homes for less. "I have set out not only to build single-family houses, but houses with their own yard, granite countertops and hardwood floors. And I'm only going to charge $85,000 to $155,000. We continue to strive to deliver more and decrease the cost per square footage."

CHAPTER 10

FILLING IN URBAN CENTERS

Urban infill — renovating blighted urban areas and creating new living areas — is one of the simplest ways of reusing our current stock of homes.

U RBAN INFILL is the use of already-developed urban areas, as opposed to the development of native land. Urban infill benefits the environment by reducing the miles people travel between their homes and area businesses, or from business to business. Densely populated cities are responsible for far less pollution than suburbs. Infill is a means of land recycling. It entails the renovation of buildings, as well as polluted city areas known as brownfields. Infill has been on the rise over the past 20 years as interest in city living has increased, but suburban development still far outweighs it, according to the publication *Urban Infill Housing Fact and Myth* by the Urban Land Institute: "After decades of losing residents, many US cities are now experiencing gains in population. Of the 20 largest cities (including only Census tracts within the city limits), 16 gained population from 1990 to 2000. In most metropolitan areas, moreover, suburban growth rates far exceeded the growth rate in the central city."[1] Infill is the reuse of entire buildings, the ultimate reclamation of building materials. As Builders of Hope exhibits, buildings can be saved in place or relocated to new city neighborhoods. Urban infill is an obvious way to reuse our stock of standing homes in need of renovation.

The Urban Land Institute (ULI) is a non-profit education and research institute supported by its members, and its stated mission is to provide responsible leadership in the use of land in order to enhance the total environment. Dedicated to helping invigorate cities and help restore their vibrancy, the ULI states in its *Infill* publication: "The realization seems to be growing that cities need good housing to become the vibrant centers of cultural and social life that they once were." The ULI supports urban infill for its contributions to city development. "Urban infill housing sparks neighborhood revitalization," the ULI contends. "Not only do new residents pay property taxes, but they also spend money. New residents spur retailing, office development, restaurant openings, cultural activities and events, religious activities, and the development of parks and recreational areas."[2]

The ULI also supports urban infill housing for its ability to make better use of limited urban land, and for its ability to financially benefit urban communities: "[Urban infill housing] reuses existing properties, which often are neighborhood eyesores, thus bringing much-needed tax dollars to local governments and revitalization to inner-city communities."

Infill also benefits the environment in numerous ways. First, it reduces the need for development of new land. "Infill development...is often less destructive to the natural environment than is suburban development," writes the ULI. "Infill housing development supports mass transit and alternative modes of transportation, including walking and biking." A study by the Northeast-Midwest Institute (NEMWI) — a Washington-based, private, non-profit and non-partisan research organization dedicated to economic vitality, environmental quality and regional equity for Northeast and Midwest states — examined the energy benefits of urban infill. It found that designing residences that are both green and efficient internally, in terms of their energy-efficiency and materials, and externally, in terms of their location relative to city services and sources of mass transit, multiplies their environmental and efficiency benefits. "This dual benefit is key," according to the NEMWI report, *Energy Benefits of Urban Infill, Brownfields, and Sustainable Urban Redevelopment*:

> Generally, green/energy-efficient buildings are designed to save about 30 percent on energy use within the structure.... Externally, "compact urban development" saves 20 to 40 percent of vehicle miles

traveled (VMT) with corresponding reductions in greenhouse gases (GHGs).... When redevelopment projects combine both elements (VMT reduction and energy-efficient buildings), the energy savings can be estimated to be 30 to 35 percent of the total energy demands attributable to the development, relative to conventional construction in suburban auto-dependent locations.[3]

But the study goes on to say that the energy effects of well-designed urban infill extend even farther, citing several other ways that urban infill enhances energy savings:

- Urban density is associated with energy efficiencies within the building due to fewer exposed surfaces. Studies indicate that multifamily buildings save between 20 and 50 percent of energy use relative to single family units (the range is largely explained by whether or not unit size is held constant).
- There is less "line-loss" in distributing electricity to dense urban areas than to spread suburban areas. Line-loss for electricity has been estimated to be nine percent of electricity production.
- Less energy is spent in building and maintaining infrastructure for urban projects than suburban sprawl projects. Limited research in this area supports a savings on the order of 25 percent attributable to urban compact development compared to suburban sprawl patterns.
- Some urban projects are served by waste-to-energy plants or district heating systems that also lower GHGs.
- An indirect benefit of urban redevelopment is the retention of greenfield "carbon sinks."
- To the extent that brownfields redevelopment involves rehabilitating existing structures instead of new construction, there is an energy savings associated with the lower energy demands of rehabilitation. While these latter factors remain insufficiently quantified (and further study is recommended), the previous point — the dual benefit of energy savings within the building envelope and VMT reduction — makes a sufficiently strong case that promotion of brownfields and sustainable urban redevelopment can be a major source of greenhouse gas reduction. This connection could

be strengthened if development incentives gave greater weight to climate change and GHG reduction. Similarly, energy policy should give greater attention to the efficiencies gained when development patterns combine the dual benefit of energy-efficient buildings in energy-efficient locations.[4]

The ULI report states that infill projects have also yielded economic and PR benefits for developers: "Developers...have discovered that although urban infill housing may be riskier, it often generates greater financial rewards than does suburban greenfield development. These projects are often high-profile developments that bring national recognition and prestige to the development company. Infill development is seen as part of the solution — not part of the problem — which is why political support for such projects is increasing in strength."[5]

Despite its many benefits and its growing number of supporters, urban infill housing is still often more difficult to build than suburban housing. Yet, the ULI says many of the difficulties of urban infill development have been blown out of proportion, and it argues that it is easier to work within existing systems than is often believed. Simple changes to existing government policy could also help encourage infill over demolition. The ULI released *Urban Infill Housing: Myth and Fact* "to clarify the misconceptions surrounding growth and development." It addresses and refutes what it calls "myths" surrounding the concept of urban infill, among them that urban land costs are prohibitive, financing for the development of urban infill projects is too complicated, and that complex city building codes and zoning laws make infill overly risky and time-consuming. The ULI says all of those myths are overblown, and that, actually, infill in urban centers is possible, economical and smart.

The NEMWI sums up its view on urban infill: "Several studies have made the connection between urban/brownfield redevelopment and the avoidance of sprawl-related environmental impacts. That is, the reuse of formerly contaminated properties located amid city neighborhoods, or infill, lessens some of the negative effects of scattered development in suburban area, or sprawl. When compared to spread-out building patterns, compact infill redevelopment produces substantial air quality and energy-related benefits."[6]

CHAPTER 11

RECLAIMING THE INNER CITY

Two landlords in Reno, Nevada, became inspired
to remodel rundown small spaces in their hometown
and ended up making a big splash.

URBAN INFILL is on the rise. The practice of reusing rundown buildings in urban cores, urban infill is the salvaging of complete structures. Institutionalizing urban infill and creating financial incentives that reward reuse over demolition help reduce architectural waste and revitalizes, rather than gentrifying, blighted urban areas.

In Reno, Nevada, HabeRae Properties is the brainchild of Kelly Rae and Pam Haberman, two landlords who became fascinated with the idea of living better with fewer resources when they built a rural getaway home for themselves outside city limits. "The impetus for building small and sustainable came to us when we bought a remote lot," Rae says. "The builder who built our little cabin—we were living in New Mexico at the time, near Mammoth—said, 'You have to go off the grid.' I said, 'What is that?' This was in 1994." Their builder, Don Berenati of Sweetwater Building, was experienced in off-the-grid, sustainable living, and he explained how the couple's home could generate its own resources. "He said, 'You can power it with solar panels; it works!" Rae relates. "And there's a spring we can tap into with a line that goes to the cistern. So I said, 'OK, I have to see that.'"

Intrigued by the idea of a completely self-sustained home, Haberman and Rae moved forward with plans to create the cabin. "We went back and forth with the fax — it was before Internet — and before I know it, we have plans for a cabin that's off the grid. A year later, there's the plan, and two years later, the house was built with an eight-panel solar system. We were amazed! All of our energy comes from these eight panels, and our water comes from this spring," Rae explains.

The cabin includes reclaimed materials, such as a salvaged wood floor from a warehouse in Oakland, California. "It was from a shipping warehouse from World War II," Rae recalls. "It was called 'the floor that won the war.'" At the same time, the pair was renovating homes in downtown Reno. As they became increasingly impressed by their sustainable cabin, they became increasingly interested in integrating some of those techniques into the homes they renovated. "We thought, 'If we can make housing better with better insulation and better materials for our own home, why can't we do this for other people and really do things with value?' It's value-engineered. I don't like the word 'cheap,' but it's economically efficient. There's a big difference," she says.

The cabin also employed a number of space-saving techniques. Haberman and Rae wanted a smaller space, and they were impressed with the efficiency Berenati worked into their little cabin. "The spaces fill more than one purpose," Haberman says. "We have window seats, and underneath are bookshelves. We can cram a lot into a small space, and then you can eliminate the need for a big house. We started doing the same thing with our properties, and we found our tenants were thinking along the same lines we were. They liked the small spaces."

The two began researching sustainable building, and they learned that for every acre of new land developed, four acres of open space are lost. They decided to change their business direction and focus exclusively on urban infill that features reclaimed building materials. "We saw what was happening here in Reno," Haberman says. "Reno is a pretty place, and we started seeing all these cookie-cutter developments. It was frustrating because that's not the city we grew up with. We started looking in the urban core and seeing there was lots we could use without going outside of city limits and cutting into the hillsides and paving everything over." Their results are wildly popular, efficient urban nests perfect for the city's large population of young college students and urban professionals. Though they have won numerous awards for historic preservation and community development, the duo says it is most important to them that the projects demonstrate how smart and invigorating reuse can be. But they also say that they are often in the position to fight city hall over rights to reuse and rebuild, rather than being encouraged to renovate rather than demolish.

Inspired by their growing knowledge of sustainable building techniques, and disappointed watching their hometown fall victim to urban sprawl, Kelly Rae and Pam Haberman started purchasing rundown urban infill properties and converting them into hip urban homes. SoDo 4 is made up of four 100-year-old, 275-square-foot brick buildings that formerly served as sleeping quarters for railroad engineers.

Pamela Haberman

■ Building Success

In 2007, Haberman and Rae found four tiny brick cottages, 100-year-old former sleeping quarters for brakemen and engineers on the old V&T Railroad, badly in need of renovation. HabeRae bought the properties, gutted layers of bad interior remodeling materials such as grungy carpet and outdated linoleum and exposed the original century-old Douglas fir floors. They tore through layers of drywall and wallpaper to uncover the original brick walls, which were left exposed in several areas. Reclaimed corrugated metal became an industrial chic ceiling. The two found an old electrical box and outfitted it as a moniker for the SoDo (short for south of downtown) homes. "It's an old electrical box we found in Unionville, Nevada, a ghost town. We put a stand on it and glued the SoDo 4 logo on it," Haberman says. "It's rusted out, but it's really cool. Instead of having something custom made, we used that — it's a piece of junk essentially." Haberman credits her partner for having the vision when it comes to reclaimed materials, for seeing the value in rundown stuff and rundown buildings: "Kelly Rae gets the credit. She's the alley cat. She gets on her bike and is pedaling up and down the street. Even I will look at it and say, 'Are you kidding me?' She'll see what it can be. It boggles my mind, but she's done it so many times."

Using inexpensive salvaged building materials allows HabeRae to offer

In the SoDo 4 homes, Haberman and Rae peeled away layers of carpeting and drywall to reveal original wood floors and brick walls. By using salvaged materials, such as corrugated metal ceilings, Haberman and Rae are able to save money; they pass along the savings to their tenants.

Pamela Haberman

a better home for less money. "If I can spend 25 cents on old metal for the ceiling, I don't spend the money I would have spent on a newly made ceiling. I don't send what's there to a landfill. I don't have to re-ceiling," Rae explains. "Then I can pass that savings on to my residents. Instead of 800, rent can be 700." HabeRae calls their small inexpensively maintained recycled homes economically efficient. "When we can do something efficient, we can pass on those savings to our residents, to our clients," Rae says. Being a small, local

For their "2 More on Watt" project, HabeRae purchased two century-old farmhouses just one mile from downtown Reno. They pulled early 1900s newspapers out of the walls, and found photographs of the Italian immigrants who once owned the land.

company with low overhead and reasonable salaries doesn't hurt, either. "We all have to make money, by why do you have to make $1 million on the project?" Rae says of her business philosophy. "Why not make $20,000 instead, as long as you have a roof over your head and a good meal on your plate? You don't have to be a millionaire, just make a decent living and places people can afford."

HabeRae says finding old stuff is easy, and that, though their residents appreciate the savings, they also just like the way old things make a home look and feel. "That old corrugated metal, there's a plethora of it," Rae says. "There's always some guy on Craigslist that had it in a barn. We use it on ceilings and walls. We don't touch it. It's rusted and gnarly and knotty, and when people walk into a house and see it on the walls or the ceiling, they're like, 'Oh God, that's cool.'"

Using salvaged materials means no two HabeRae homes are identical. Reclaimed materials bring their own history and story to each project. Rae loves how using old buildings and old materials gives her projects a sense of history: "Buildings have history, and I believe they talk to us. They have the wonderful stories of all the people who lived there, and when a new owner comes in, they start their own story." She references one of their renovation projects, where they uncovered parts of the region's history: "Take the old farmhouse on Watt Street. We pulled newspapers from 1902 out of those walls — they were used as insulation. They were newspapers from Italy. All of that area used to be farmland, and Italian immigrants made their home in Reno to farm. That's a story to be told, and from this point forward, the people in this building make their own history."

HabeRae adds other touches to make their developments homes, not just houses. They incorporate garden spaces where tenants can grow their own food. "We provide the place, a drip system, everything," Rae explains. "It's set up before they move in." The homes also emphasize access to mass transit and car-free living. "All of our projects are within five minutes by bicycle to downtown Reno or to the university. They're all on a bus line," Rae says. "You don't need a car." Haberman says their practices attract great residents, many of them young people who are attracted to their properties' small size and historic feel: "We get much better tenants because they see that we care. They want to live somewhere where the landlord cares."

Haberman and Rae put a lot of thought into the end user when designing their remodels. They want to create homes that are great to live in and affordable, but also well-designed. "I think what sets us apart in our city is that we know what counts to people," Haberman maintains. "It has a custom or designer feel to it without it costing a lot of money. We do a lot of unique, hip things. In one place, we took stucco off the walls and exposed the original brick. We reclaimed the flooring — there was carpet, linoleum, adhesive — we took off layer after layer of that stuff to get to the original Douglas fir floors." She says designing small spaces well also shows people what's possible in a

A classic example of urban infill, HabeRae's "2 on Watt" project was built on an abandoned lot in an old part of Reno. The cottage style was designed to fit in with the homes' historic surroundings.

Pamela Haberman

reduced amount of space: "We have everything self-contained in a very small space. We include everything a person could need — a stacked washer and dryer, a dishwasher — everything a person needs as a tenant. And it's all in this compact, tight space. Everyone is just clamoring for simplicity, and they don't always know how to go about getting it. We're setting the example, and people realize you can live like that. You can live a smaller, less complicated life."

■ Hometown Care

HabeRae states that their homes make a positive impact on their community by bringing new life to formerly blighted areas, and by providing affordable homes residents are proud of: "We're always working in the dead areas of

The 2 on Cheney project includes a 1930s Spanish-style bungalow on a front lot and a cottage on a back lot. Both were restored using much of their original materials.

town," Haberman says. "I don't know why, we're just attracted to them. The places everyone says you should stay out of, we're the first ones there." She loves how their projects become the spark that helps bring life back to dead areas of town. "The next time people come, it's a whole new area in what was just a blighted area of town. You clean it up, give it some attention, and it attracts people back."

The duo also loves seeing how their efforts encourage community among the homeowners who move in. "When we first finished SoDo 4, the gal that moved into one of them had a huge housewarming party. There were 40 people there, in a 275-square-foot house! They spilled onto the deck," Rae recalls. She says revitalized neighborhoods where residents feel safe changes

HabeRae includes amenities such as built-in garden beds that make living the simple life easy in their homes.

Pamela Haberman

the dynamic of an area: "It encourages community. People aren't so inclined to press the button, go into the garage and press another button when they're proud of where they live. They want to tell people about it. They want to share their joy."

HabeRae also believes their developments support the philosophy of wise urban planning and mixed-use development. They view the renovation of older buildings as vital to communities not just because of the benefits to homeowners, but also because it economically benefits the entire community. Mixing lower- and higher-cost real estate in a community allows for healthy business development, Rae says: "Every community has to have a mix of new and old. The landowners of the older places probably owe less on their mortgage, so the landowners that own those places can charge less in rent from a prospective business. Let's say they can charge 50 cents in rent. Developers

HabeRae rescued an abandoned 1950s firehouse in a dangerous part of the city. Now a 5,000-square-foot mixed-use building, the space offers a bagel shop, a salon and nine residential lofts.

Pamela Haberman

who build these new buildings have to charge $1.50, but the small business owner can't afford $1.50. You have to give those small business owners a chance to build a company at reduced-rate rent. When they become Microsoft, they can go pay the big rent."

Kelly Rae and Pam Haberman think the key to their business's success is that, by reusing old spaces and providing smaller, easier-to-maintain hous-ing, they appeal to people's desires for a simpler life. Rae says her renters and homebuyers "want to live a simple life that uses less energy."

"They're putting a lot into their jobs and their education trying to make it in this society," Rae says. "They don't want to be bothered with a large house, a large payment, a large anything." As they move forward with their projects, Haberman and Rae rely on their own feelings and instincts — when they think something is cool, others seem to as well. "Pam and I recognized this trend within us," Rae says. "We just said, 'We're tired of this. We want to move into somewhere small.' That's how we want to live. We don't want to have to worry about a big place, a big this, a big that. And if we like it, we think other people will like it. We've been pretty lucky that way so far."

Haberman says having a smaller home that costs less in terms of money and maintenance has benefits that ex-tend beyond homeowners' wallets. She sees living in a smaller home as the first

The firehouse lofts include original concrete floors, custom glass mosaic tile work, sleeping lofts, high-efficiency heating and cooling systems and washers and dryers. A rooftop deck offers outdoor spaces with organic garden beds and drought-tolerant urban landscaping.

step toward living a life that values relationships over things: "I think every-body's just tired of the pace of their lives, of the amount of hours they have to work and of being tied to a big house. They have to spend a lot of money just putting stuff in it. It takes a lot of time to maintain it. It makes you a slave to your house, and your actual life goes away. Your kids are being cared for by someone else because you have to work so much to pay for your house."

All the firehouse lofts include massive windows; residence 9's roll-up wall of glass used to be the fire truck door.

Fighting the Good Fight

Replicating the HabeRae model makes sense for landowners and renovators. Landowners who consider reuse can save on building supplies, and salvaging architectural materials makes for a lower-cost overhaul. However, though reusing old buildings seems to benefit the community and the environment, HabeRae says they often find themselves fighting city hall rather than working with them.

Haberman laments a fight with city hall that she and Rae lost: "There was an old gospel mission in downtown Reno — a beautiful old building with Spanish tile floor, stucco arches. It was an old Spanish mission." The duo had recently finished another renovation project, and she says, "We were hot to trot to do another old building." They heard the mission was scheduled for demolition, so they tried to intervene. They called the real estate manager. "It took three or four phone calls, but finally someone gets back to us. We wanted to go look at it and they said, 'Well, you have to wear a hardhat. It's about ready to fall down,' blah, blah. We went and looked, and in our minds, it was in great condition. It was a beautiful old historic building, and at the thought of tearing it down, we were like, 'You've got to be kidding me!'" The owners of the property also owned several area casinos, and they planned to tear down the mission to build a parking lot. "We met with the city council and said, 'Before you tear it down, let us take a shot at it,'" Haberman says. Though the city council said they would consider the offer, they ended up allowing the owner to demolish the old building. "Now, instead of this beautiful old building, we have a parking lot that's almost always empty. They're all about tearing down the old and making it new, but that's so boring," Haberman states. "It's like taking all the senior citizens out and shooting them! You would never do that. Their experience enriches everybody's life. It's the same with old buildings."

Despite disappointments, HabeRae uses their projects as a tool to propel their city forward, and say others could do the same. "It starts with local government," Rae explains. "We've really had to educate our community and the development section of the city of Reno." She laments that often reuse seems to come with additional hurdles rather than aid: "It's too bad there is no incentive to save old buildings. In fact, it's the opposite. If you want to save an old building or get historical status, it's going to cost you. You have to have a passion for it. You have to have intestinal fortitude. You can get discouraged, and, oh boy, you just want to go to a beach." But Haberman and Rae have proven to the City of Reno that their developments work, and they've broken through many barriers with their innovative projects. "You have to show them examples. You can't say, 'Hey, I know everything, believe what I say,'" Rae says. "You have to show them a better way. You have to be the leader for change."

And though changing minds in city hall is satisfying, changing homes and lives is HabeRae's goal. "Life energy is precious, and we don't want to spend our life energy on big houses and big yards," Rae says. "We want to live a simple life. I think that's important and brings a more social, communal country and community. I think that's definitely lacking, and I hope we can change it."

CHAPTER 12

SALVAGING
SMART CITIES

Improving our cities' efficiency, providing quality
low-income urban housing and increasing safety and
cooperation in urban centers are vital as we move into the
future. By creating a nationwide system of deconstructing
our current housing stock and cataloguing and reusing
the supplies, we can decrease the cost of building and
remodeling, and revitalize urban communities.

S MART, SUSTAINABLE urban development is a necessity. Not a political
issue, sustainability in our city development is the way of the future. By
choosing not to improve our urban livability by increasing walkability, community, safety and affordable housing, we are allowing ourselves to fall behind
nations like Brazil and Japan in terms of efficiency and sustainability, which
will translate into an economic deficiency as energy prices peak in coming
years.

Designing sustainable cities that encourage community is a crucial part
of the sustainability of our species moving into the future. Wisely using our
reclaimed building supplies and housing stock can work into a sustainable
vision of our cities. Urban planners are already working toward, and especially
in other countries, implementing ways to improve the ecology of our modern
cities, redesigning urban transport, encouraging walking and biking, recycling
water, incorporating farming and providing higher-quality low-income housing settlements. In his book *Plan B 4.0*,[1] Lester Brown, president of the research organization Earth Policy Institute in Washington DC, outlines a plan

for a sustainable future. His elegant, brilliant solution involves redesigning urban centers. Let's consider how reclaimed housing can be integrated into that solution.

■ The Importance of Cities

First, it's important to understand the vitality of redesigning cities. As of 2008, for the first time in history, more than half of human beings live in urban settings. In the past 100 years, we have rapidly shifted from an agriculture-based species to an urban-based one. In 1900, 150 million people lived in cities. In 2000, 2.8 billion lived in cities — a 19-fold increase, Brown writes. He also notes our rapidly expanding city populations: In 1900, a handful of cities worldwide had a population of a million people or more. Today, 431 cities globally have that many inhabitants; 19 megacities — among them New York, Tokyo, Mexico City, Mumbai, Delhi, Shanghai, Sao Paulo and Beijing — have more than 10 million residents. In many of these cities, redesigning isn't an option, it is a necessity as pollution renders air unbreatheable and congestion causes disruptive transportation delays. The matter is also economic. The Partnership for New York City, a group representing New York's leading corporate and investment firms, estimates that traffic congestion in its metropolitan area costs the region $13 billion a year in lost time and productivity, wasted fuel and lost business revenue, Brown reports.

Despite the congestion and pollution hurdles many cities face, densely populated cities are beneficial from an environmental perspective. Compared with suburbanites, city dwellers use public transportation more, take advantage of densely packed businesses and often participate in green initiatives more. A 2008 study published by the Brookings Institute, a non-profit public policy organization in Washington DC, reported that the average American living in a metropolitan area had a carbon footprint 14 percent lower than those who lived outside cities.[2] Other studies support the findings — the *New York Times* reported on several in its April 2009 article "The Relatively Green Urban Jungle": A study by Harvard economics professor Edward L. Glaeser and UCLA environmental ecologist Matthew Kahn found that cities generally have lower carbon outputs than suburbs. A worldwide study by David

Dodman of the International Institute for Environment and Development in London found that per capita emissions from 12 major cities in Europe, Asia and North and South America were typically lower than their national average of per capita emissions. A study in China found that not only do city dwellers emit less carbon, but they are also more likely to participate in environmentally beneficial activities, reports *Discovery News*.[3] The study, led by Michigan State University sustainability scientist Jianguo "Jack" Liu, surveyed 5,000 Chinese residents living in small and huge cities. The questions asked if, in the last year, residents had sorted garbage, talked about environmental issues, recycled plastic packaging bags, participated in environmental education programs or been involved with environmental litigation. The study found the environmentally active behaviors were more common among residents in the biggest cities than in residents in small cities. The researchers' analysis also showed that environmentalism wasn't linked to wealth or incomes — it was linked to employment, probably because many Chinese companies and organizations encourage employees to take environmental action by doing things like planting trees together, according to *Discovery News*.

City municipalities worldwide have demonstrated how to encourage shifts toward more environmentally beneficial behaviors through tax and transportation policies. Some cities have encouraged a shift away from car culture, which emits much more carbon than mass transit and biking and walking, by charging cars to enter cities. Singapore, Oslo, London and Stockholm are among major cities who charge congestion fees to vehicles entering the city. London adopted a congestion fee in early 2003, charging vehicles entering the city between 7 a.m. and 6:30 p.m. a fee of £5 (roughly $8 at the time), reducing the number of cars traveling on city streets during peak times, Brown writes.[4] In its first year, the tax led to a 38 percent increase in people taking buses into central London, and increased speed on key thoroughfares by 21 percent. The tax was increased and expanded in 2005, with the revenue being used to increase and improve public transportation. Since the adoption of the congestion fees, the daily flow of cars and cabs into central London during peak hours has dropped by 36 percent, while the number of bicycles has increased 66 percent, Brown reports. In Paris, government officials are working to reduce traffic congestion, with a goal of a 40 percent reduction by 2020. To support the goal, the City has adopted both congestion fees and a city bike rental program, which offers more than 20,000 bicycles for rent at 1,450 docking stations. The popular program is being mimicked by cities and universities in the United States. St. Xavier University in Chicago and Emory University in Atlanta have adopted bike-sharing programs to reduce traffic congestion on campus. At the University of New England in Maine and Ripon College in Wisconsin, incoming freshmen receive a free bike if they agree to leave their cars at home, Brown reports, a move the colleges say saves money over the cost of providing parking and congestion management.

All of these traffic-reduction methods showcase the ways in which cities can actively participate in a shift of resident behavior toward environmental, community and health-enhancing choices. Brown cites Bogota, Colombia, as one of the most rapidly altered cities in the world. Mayor Enrique Penalosa served as mayor of Bogota for three years, taking office in 1998. Penalosa was devoted to improving life for the 70 percent of his citizens who didn't travel by car. In his short tenure, he "banned the parking of cars on sidewalks, created or renovated 1,200 parks, introduced a highly successful bus-based rapid transit system, built hundreds of kilometers of bicycle paths and pedestrian streets, reduced rush hour traffic by 40 percent, planted 100,000 trees, and involved local citizens directly in the improvement of their neighborhoods," Brown writes.[5]

■ Reclaiming the City

How does reclaimed housing work to help shift urban environments toward sustainability? As we've examined via the Phoenix Commotion, Builders of Hope and HabeRae, saving and improving urban homes increases civic pride and involvement in communities and cities. By reclaiming abandoned urban properties and structures, we increase the density of residential use of the city, allowing more people to live well in cities, rather than suburbs, and to live with nearby access to mass transit, along with jobs and businesses. Reclaiming rundown urban spaces also helps improve civic pride, which studies show decreases crime and fear of crime. Numerous studies and real-life examples show how renovating and using rundown city spaces improve city life. A community garden planted in Kitchener, Ontario, Canada, on the site of a former dump reduced crime by 30 percent in its first summer, reports the London Commission for Architecture and the Built Environment (CABE).[6] In Los Angeles, a blighted park was renovated via a Summer Night Lights program. Municipal lights are left on in the park past midnight, and the city program offers free food, dancing lessons and sport leagues, along with increased law enforcement and gang intervention workers. In its first summer, the program saw a 17 percent decrease in violent, gang-related crimes around the parks, and a 10 percent overall city decrease in crime, reports the *Wall Street Journal*.[7]

Increasing home ownership also enhances neighborhood safety and community involvement. A study conducted by Robert Dietz of the Ohio State University Department of Economics and Center for Regional and Urban Analysis and funded by the Homeownership Alliance found that "through their investment in the home — and therefore in the local neighborhood — homeowners appear to be overall more involved in their communities. These efforts by homeowners generate benefits for their communities in addition to the benefits for their families." Dietz cites four major benefits to homeowners and communities by increased home ownership:

1. Children of homeowners are likely to perform higher on academic achievement tests and are more likely to finish high school. Furthermore, children of homeowners have fewer behavioral problems in school and are less likely to become pregnant as teenagers. These outcomes survive many controls for parental education, marital status, and other statistical comparisons, as well as neighborhood characteristics.

2. Political activity, like voting, as well as participation in civic organizations is higher among homeowners than renters after controlling for personal characteristics and socioeconomic status.

3. Homeowners, again once controls are in place, are more satisfied with their lives and are happier.

4. Some of the most recent research suggests that a high level of homeownership in neighborhoods enhances property values.[8]

If densely packed cities, with initiatives to improve transportation habits, and increased homeownership work together to improve urban safety and community involvement, using our vast stock of rundown, impoverished and abandoned housing creates a way to tie together these two goals. We can use inexpensive or free materials to build low-income housing rather than sending them to the landfill. Studies also show that when individuals directly participate in the improvement of their communities, it increases their investment in that community. Reusing and refurbishing urban buildings provides the opportunity to get community members involved in the betterment of communities. Taking inspiration from the Phoenix Commotion, an unskilled labor force that includes future homeowners could be used to build and renovate homes, providing training to these workers. Combining the savings of free or low-cost materials and labor will allow these reclaimed homes to be affordable for current city residents, improving communities without forcing out current residents.

By instituting a nationwide program that oversees the deconstruction of homes and catalogs and stores building supplies, we can create a stock of materials that renovators could depend upon to renovate homes and businesses in blighted urban areas. By creating a team of workers, with professional builders overseeing a rotating crew of untrained employees, who are tasked with renovating neighborhoods and homes, we can improve our cities by increasing our number of affordable homes and, therefore, our number of engaged, low-income homeowners.

We could also utilize our stock of used housing to improve conditions for the very poor and homeless. According to Brown, "Between 2000 and 2050, world population is projected to grow by 3 billion, but little of this growth is projected for industrial countries or for the rural developing world. Nearly all of it will take place in cities in developing countries, with much of the urban growth taking place in squatter settlements."[9] Squatter settlements the world over are inhabited by the extremely poor and are characterized by "grossly inadequate" housing and a lack of access to urban services, Brown writes. Most are extremely impoverished migrants who have moved from rural to urban settings in an attempt to improve their access to opportunities. Brown suggests one way to decrease squatter dwellings is to increase spending in smaller cities and rural areas in an attempt to curb migration to urban centers. But squatter settlements will continue to exist and grow; to address them adequately, cities must support squatter settlements rather than attempt to eliminate them. Curitiba in Brazil is on the cutting edge of urban planning. The City designated specific tracts of land for squatter settlements, allowing the City to better structure the design of the areas and to improve basic provisions such as clean water and healthy waste disposal, decreasing disease and its associated medical expenses. Regular bus services also offer transportation to places of employment, Brown writes.

Brown suggests planning these communities so they are places where individuals are fostered to improve their lives. By incorporating community spaces and parks, we can decrease crime and improve safety and togetherness in these spaces. By providing transportation, we can increase employment opportunities. As Brown contends, "Although political leaders might hope these settlements will one day be abandoned, the reality is that they will continue expanding. The challenge is to integrate them into urban life in a humane and organized way that provides hope through the potential for upgrading. The alternative is mounting resentment, social friction and violence."[10]

If our ability to improve society as a whole depends on our ability to provide adequate, safe, healthy very low-income housing, the supply of used housing materials can play a big role. A nationwide program that deconstructs and stores building materials could not only provide supplies for low-income urban communities, but also provide them for a stock of settlements that aid the destitute. Decreasing the cost of building supplies for these homes helps make government assistance programs of this sort more affordable, and reusing materials in this way keeps them out of the landfill, doing good rather than languishing.

CONCLUSION

THE PEOPLE, HOMES AND ORGANIZATIONS in this book show that creating a home can mean so much more than moving into a nondescript vinyl box in the suburbs. Across the nation, these people and many, many more are showcasing alternatives to the standard housing model. These people reject the standard model, and they forge a new path toward homeownership. Amazingly, they have all done it with little to no debt, creating affordable, unique homes and reusing waste in the process.

We send millions of pounds of demolition waste to the landfill every year. Our standard model today is the laziest and most wasteful possible model — completely demolishing homes with perfectly reusable, high-quality materials inside and adding them to giant mounds of waste around the world. We must change this model.

In this book, I hoped to explore some of the many options for reusing high-quality building materials. Whether the materials are lovingly collected one at a time in towns all over one county in Alabama or delivered by the truckload to a warehouse in Texas, we know they can be put to good use because good people all over the place are already doing it. Though many organizations are attempting to bring attention to this subject, it will require a change in protocol throughout the construction/deconstruction industry to truly change the status quo.

Many models exist that show how we could institutionalize the dismantling of buildings and the repurposing of the materials inside. The examples in this book are far from exhaustive. Everywhere across the nation, individuals make the choice to deconstruct buildings responsibly, to recycle the supplies within, to gather used items rather than buying new. We all should be doing things their way.

Though all recycling is valuable, recycling housing materials offers more than just waste reduction; it offers the almost incalculable value of providing affordable homes in cities across the nation. Reusing housing materials can positively affect our economy. It can positively affect our communities. It can be a key tool in providing high-quality, healthy and affordable housing to the many families and individuals in need. As countless studies show, more healthy homes and invested homeowners means more community engagement. Less crime. Healthier kids. Improved graduation rates.

The tide is turning toward reuse. Executive Order 13514, signed by President Obama in October 2009, requires that all federal agencies divert at least 50 percent of their non-hazardous solid waste by the end of 2015. We need this law to apply to every structure in the United States. There is much we can do to implement change in this arena. Contact your local government agencies in charge of waste management and ask about their deconstruction reuse efforts. Find out if you can help or get involved, and help spread the word about deconstruction options in your community. Habitat for Humanity ReStores nationwide accept used building materials then resell them. Consider volunteering. Get your local government to care about reuse. Get your state government to care. Or go out and collect some salvage and build your own house. You'll get a home out of it.

Notes

Introduction

1. nrel.gov/analysis/seminar/docs/ea_seminar_aug_12.ppt, accessed March 5, 2011.
2. epa.gov/region9/waste/solid/stardust/index.html, accessed March 5, 2011.
3. nlihc.org/doc/Prelim-Assess-Rental-Affordability-Gap-State-Level-ACS-12-01.pdf, accessed March 5, 2011.

Chapter 2

1. corelogic.com/About-Us/News/New-CoreLogic-Data-Shows-Second-Consecutive-Quarterly-Decline-in-Negative-Equity.aspx.
2. Anthony J. Badger, *The New Deal: The Depression Years, 1933–1940*, p. 239.
3. archives.gov/education/lessons/homestead-act.
4. urbansculpture.com/history.
5. thehistoryof.net/history-of-home-mortgages.html.
6. howstuffworks.com/framed.htm?parent=prefab-house.htm&url=searsarchives.com/homes.
7. grow-management.Alachua.fl.us./building/build code.php.
8. Badger, *The New Deal*, p. 239.
9. Ibid.
10. Ibid., p. 242.
11. "Who Can Afford to Live in a Home: A Look at Data from the 2006 American Community Survey," prepared for the US Census Bureau.
12. census.gov/hhes/www/housing/special-topics/files/who-can-afford.pdf.

Chapter 4

1. Robert Shiller, *The Subprime Solution*, p. 45.
2. nytimes.com/2008/12/21/business/worldbusiness/21iht-admin.4.18853088.html?pagewanted=1&_r=1, accessed January 10, 2011.
3. Robert Shiller, *Irrational Exuberance*, p. 38.
4. nytimes.com/2008/12/21/business/worldbusiness/21iht-admin.4.18853088.html?pagewanted=2&_r=1, accessed January 10, 2011.
5. *The Giant Pool of Money*, *This American Life*, NPR, originally aired May 9, 2008; thisamericanlife.org/radio-archives/episode/355/the-giant-pool-of-money.
6. Jo Becker, Sheryl Gay Stolberg and Stephen Labaton, "Bush Drive for Home Ownership Fueled Housing Crisis," *New York Times*, December 21, 2008.

7. asanet.org/press/foreclosure_crisis_and_race.cfm, accessed December 14, 2010.

8. CNNMoney.com, accessed January 10, 2011.

9. Shiller, *The Subprime Solution*, pp. 71–72.

10. Ibid., p. 69.

11. Ibid., p. 73.

12. National Renewable Energy Laboratory, nrel.gov/analysis/seminar/docs/ea_seminar_aug_12.ppt.

13. Douglas Farr, *Sustainable Urbanism: Urban Design with Nature*, p. 367.

14. Shiller, *The Subprime Solution*, pp. 76–77.

15. Ibid.

16. babcockranchflorida.com/, accessed January 10, 2011.

17. housingworksri.org/, homepage as of March 31, 2011.

18. nhc.org/media/documents/FramingIssues_Heath.pdf.

Chapter 6

1. earthship.org/index.php?option=com_content&view=article&id=356, accessed April 2, 2011.

2. energystar.gov/index.cfm?c=home_sealing.hm_improvement_insulation_table, accessed December 2010.

3. See naturalhomemagazine.com/energy-efficiency/plant-your-way-to-energy-savings-landscaping-energy-efficiency.aspx, accessed December 2010.

Chapter 9

1. buildersofhope.org.

Chapter 10

1. uli.org/ResearchAndPublications/Reports/~/media/Documents/ResearchAndPublications/Reports/Affordable%20Housing/Urban%20Infill.ashx, accessed April 4, 2011.

2. Ibid., p. 4.

3. nemw.org/images/stories/documents/energy_benefits_infill_brfds_final_12-08.pdf, p. 3, accessed April 4, 2011.

4. Ibid.

5. uli.org/ResearchAndPublications/Reports/~/media/Documents/ResearchAndPublications/Reports/Affordable%20Housing/Urban%20Infill.ashx, p. 4, accessed April 4, 2011.

6. nemw.org/images/stories/documents/energy_benefits_infill_brfds_final_12-08.pdf, p. 3, accessed April 4, 2011.

Chapter 12

1. Lester R. Brown, *Plan B 4.0: Mobilizing to Save Civilization.*
2. brookings.edu/~/media/Files/rc/papers/2008/05_carbon_footprint_sarzynski /carbonfootprint_brief.pdf, accessed January 2011.
3. news.discovery.com/earth/urban-green-china-environment-110118.html, accessed January 2011.
4. Brown, pp. 197–199.
5. Ibid., pp. 193–194.
6. webarchive.nationalarchives.gov.uk/20110118095356/www.cabe.org.uk/files/the -value-of-good-design.pdf, part F, accessed April 4, 2011.
7. online.wsj.com/article/SB10001424052748704249004575385754270183396.html, accessed January 2011.
8. newtowncdc.org/pdf/social_consequences_study.pdf, accessed January 2011.
9. Brown, p. 160.
10. Ibid., p. 162.

Resources

Government Associations and Non-profits Dedicated to Increasing Use of Deconstruction and Architectural Salvage

Nationwide, organizations dedicated to reuse, deconstruction and waste management work with the public and government institutions to increase awareness and popularity of deconstruction and reclamation.

Nationwide

Building Materials Reuse Association, bmra.org

ReConnX, develops tools that support the deconstruction industry, nailkicker.com, Boulder, CO, 888-447-3873

The Reuse People, online with offices nationwide, thereusepeople.org, 510-383-1983, building deconstruction, building materials salvage, building materials distribution, project management, training, reuse and recycling plans

Waste to Wealth, ilsr.org/recycling/decon/, Minneapolis, Washington DC and projects nationwide, DC: 202-898-1610; MN: 612-379-3815

Alaska

Department of Conservation/Valley Materials Exchange, Big Lake, 907-892-7188

California

California League of Conservation Voters, Los Angeles and Oakland, ecovote.org, 800-755-3224

Deconstruction and Reuse Network, Los Angeles and Orange County, serving sites statewide, reusenetwork.org, 888-545-8333

Connecticut

Department of Environmental Protection, Source Reduction and Recycling Program, Bureau of Materials Management and Compliance Assurance, Hartford, ct.gov/dep, 860-424-3365

Georgia

EnviroShare Materials Exchange, Gainesville, hallcounty.org/enviroshare, 770-535-8284

Hawaii

Re-use Hawai'I, Honolulu, reusehawaii.org

Kentucky

Bluegrass Regional Recycling Corporation, thebrrc.com, 859-626-9117

Massachusetts

Massachusetts Department of Environmental Protection, Waste Reduction Programs, Boston, mass.gov/dep/recycle, 617-292-5500

Minnesota

Minnesota Resource Recovery Association, statewide, mnresourcerecovery.com, 651-222-7227
The Reuse Center, Maplewood, thereusecenter.com, 651-379-1280

New Hampshire

New Hampshire Department of Environmental Services, Waste Management Division, Concord, des.nh.gov, 603-271-3503

New York

New York State Department of Environmental Conservation, Bureau of Waste Reduction and Recycling, Albany, dec.ny.gov, 518-402-8678

Northeast

Northeast Recycling Council, Brattleboro, Vermont, nerc.org, 802-254-3636

Rhode Island

Rhode Island Resource Recovery Corporation, Johnston, rirrc.org, 401-942-1430

Vermont

Vermont Agency of Natural Resources, Department of Environmental Conservation, Waste Management Division, Waterbury, anr.state.vt.us/dec/wastediv/index.htm

Washington

King County Solid Waste Division, Seattle, kingcounty.gov, 800-325-6165 ext. 64466

Selective Architectural Salvage and Antique Stores

Though more expensive than seeking salvage bound for landfills, architectural salvage stores offer treasures, and prices vary. Some listed here are high-end, while others offer quite reasonable prices. All offer inspiration and a taste of the beauty of the past. For those looking for specific items, or those who are happy to pay for someone else's research and legwork, these stores offer a great resource.

Alabama

Southern Accents Architectural Antiques, Cullman, sa1969.com, 877-737-0554

California

Amighini Architectural, Anaheim, salvageantiques.com, 714-776-5555

Antique Building Materials Company, Ben Lomond, business nationwide, architectural auction.com, 415-565-0287

Corona del Mar, 949-673-5555

Architectural Antique & Salvage Co., Santa Barbara, 805-965-2446

Architectural Salvage of San Diego, San Diego, architecturalsalvagesd.com, 619-696-1313

Gayle's Pasadena Architectural Salvage, Pasadena, pasadenaarchitecturalsalvage.com, 626-535-9655

Ohmega Salvage, Berkeley, ohmegasalvage.com, 510-843-7368

Olde Good Things, Los Angeles, ogtstore.com, 213-746-8600

Tony's Architectural Salvage, Orange, 714-538-1900

Vintage Timberworks, Temecula, vintagetimber.com, 951-695-1003; West Los Angeles, 310-477-8444

Colorado

Architectural Salvage Inc./Salvage Lady, Denver, salvagelady.com, 303-321-0200

ReSource Reclaimed Building Materials, Boulder, resourceyard.org, 303-419-5418

Florida

Allison's Adam & Eve Architectural Salvage Company, West Palm Beach, adamandeve salvage.com, 561-655-1022

Architectural Antiques, Miami, miamiantique.com, 305-285-1330

Architectural Design and Artifacts, Fort Lauderdale, 954-525-1212

Sarasota Architectural Salvage, Sarasota, sarasotasalvage.com, 941-362-0803

Georgia

Metropolitan Artifacts, Atlanta, 707-986-0007

Restorations and Antique Supplies, Savannah, 912-236-7724

Iowa

West End Architectural Salvage, Des Moines, westendarchsalvage.com, 515-243-0499

Indiana

Doc's Architectural Salvage and Reclamation Services, Indianapolis, docsarchitectural salvage.com, 317-924-4000

Illinois

Spiess Architectural Antiques, Joliet, 815-722-5639

Kentucky

Architectural Salvage, Louisville, architecturalsalvage.com, 502-589-0670

Maine

The Old House Parts Company, Kennebunk, oldhouseparts.com, 207-985-1999
Portland Architectural Salvage, Portland, portlandsalvage.com, 207-780-0634

Michigan

Architectural Salvage Warehouse of Detroit, Detroit, aswdetroit.org, 313-896-8333

Minnesota

Architectural Antiques, Minneapolis, archantiques.com, 612-332-8344
City Salvage, Minneapolis, citysalvage.com, 612-627-9107

Missouri

Antiquities and Oddities Architectural Salvage, Kansas City, aoarchitecturalsalvage .com, 816-283-3740
Urban Mining Homewares, Kansas City, urbanmininghomewares.com, 816-529-2829

New Hampshire

Architectural Salvage Inc., Exeter, oldhousessalvage.com, 603-773-5635
Nor'East Architectural Antiques, South Hampton, noreast1.com, 603-394-0006

New Jersey

Amighini Architectural Salvage, salvageantiques.com, 201-222-6367
Recycling the Past, Barnegat, recyclingthepast.com, 609-660-9790

New York

Historic Houseparts, Rochester, historichouseparts.com, 585-325-3613
Olde Good Things, Manhattan, ogtstore.com, two locations, 212-989-8814; 888-551-7333
Upper West Side, 212-362-8025

Ohio

Columbus Architectural Salvage, Columbus, columbusarchitecturalsalvage.com, 614-299-6627

Oregon

Aurora Mills Architectural Salvage, Aurora, auroramills.com, 503-678-6083

Pennsylvania

Old Goode Things, Scranton, ogtstore.com, 888-233-9678

Provenance Old Soul Architectural Salvage, Philadelphia, phillyprovenance.com, 215-925-2002

Texas

Discount Home Warehouse, Dallas, dhwsalvage.com, 214-631-2755

Vermont

Architectural Salvage Warehouse/Mason Brothers Architecturals and Antiques, Essex Junction, architecturalsalvagevt.com, 802-879-4221

Discovery Architectural Antiques, Gonzalez, discoverys.net, 830-672-2428

Virginia

Black Dog Salvage, Roanoke, blackdogsalvage.com, 540-343-6200

Caravati's Inc. Architectural Salvage, Richmond, caravatis.com, 804-232-4175

Washington

Earthwise Architectural Salvage, Seattle, earthwise-salvage.com, 206-624-4510

Materials Exchange/ReStores/Non-profits

Nationwide

ReStores, national retailers, habitat.org/restores

Recycler's World, online exchange, recycle.net

Alabama

Dixie Salvage, Fort Payne, 256-845-5475

James and Company Antique Timbers and Flooring, jamesandcompany.com, 256-997-0703

Elrod's Building Material and Salvage, Bessemer, 205-426-8788

Surplus and Salvage Sales, Birmingham, 205-592-8306

Alaska

Alaska Materials Exchange, online, 907-269-7586

Arizona

Gerson's Used Building Materials, Tucson, gersons.net, 520-624-8585
Salvage Depot Inc., Glendale, recyclearizona.com, 623-680-4874
Taylor Demolition and Recycling, Tucson, 520-623-0410

Arkansas

Architectural Salvage by Ri-Jo, Mena, 479-394-2438
Bear's Building Salvage, Fayetteville, 501-443-2327

California

C & K Salvage, Oakland, 510-569-2070
Caldwell Building Salvage Resource, San Francisco, 415-550 6777
California Materials Exchange (CalMax), statewide, calrecycle.ca.gov, 916-322-4027
Castroville Used Building Materials, Castroville, 415-550 6777
Cornerstone Salvage Company, Littleriver, 707-937-5011
Delta Scrap, Salvage & Demo, Oakley, deltascrap.com, 925-754-1474
Garbage Reincarnation Inc., Santa Rosa, garbage.org, 707-795-3660
Freeway Building Materials, Los Angeles, freewaybuildingmaterials.us, 323-261-8904
Nunez and Sons Used Building Material, Los Angeles, 323-266-0518
Terra Mai, Mt. Shasta, terramai.com, 530-926.6100
Whole House Building Supply & Salvage, San Mateo, driftwoodsalvage.com,
   650-558-1400

Colorado

Alpine Bargain Center, Loveland, 970-622-8307
Oxford Recycling Inc., Englewood, oxfordrecycling.com, 303-762-1160

Connecticut

The ReCONNstruction Center, New Britain, reconnstructioncenter.org, 860-597-3390

Florida

Builders Bargain Surplus, Fort Lauderdale, builders-bargain.com, 954-522-1900
Discount Building Material, Daytona Beach, 386-255-0002
Florida Wrecking and Salvage, Gibsonton, 813-741-0405
Globe Demolition and Recycling, Orlando, 407-422-4768
L&L Demolition and Salvage, Orlando, 407-295-0875

Pensacola Salvage 7, Pensacola, 850-455-7000

ReUser Building Products, Gainesville, 352-379-4600

Roz's Reusable Building Materials, Port Charlotte, 941-766-0004

Used Stuff Inc., Sarasota, 941-953-5100

Georgia

Atlanta Salvage Outlet, Atlanta, 404-873-4416

Builder's Salvage, Rome, 706-232-8869

EnviroShare Materials Exchange, Gainesville, hallcounty.org/enviroshare, 770-535-8284

Home Resource Interchange, Atlanta, 404-624-4434

Northside Material Brokers, Atlanta, northsidematerialbrokers.com, 404-762-9906

ReUse the Past Inc., Grantville, 770-583-3111

Wrecking Barn, Atlanta, 404-525-0468

Idaho

Building Material Thrift Store, Hailey, buildingmaterialthriftstore.org, 208-788-0014

Illinois

Builders Salvage, Farmer City, 309-928-2344

Lowder Construction Architectural Salvage, Waverly, 217-435-9618

Mid-America Architectural Salvage, Grayslake, 847-223-5722

PlanetReuse, Chicago (also in Kansas City), planetreuse.com, 816-686-8377

The Reuse People of America, Northbrook, thereusepeople.org, 510-383-1983

Salvage One, Chicago, salvageone.com, 312-733-0098

Indiana

Capellier Salvage and Wreckage, Camby, 317-831-4533

First Saturday Construction Salvage, Spencer, 812-876-6347

Rehab Resource Inc., Indianapolis, rehabresource.org, 800-685-4686

Richey Salvage and Demolition, Greensburg, richeysalvage.com, 812-663-6512

Tim & Billy's Salvage Store, Indianapolis, 317-632-7161

Tim and Avi's Salvage Store, Indianapolis, 317-925-6071

Iowa

Cedar Valley Recovery and Demolition, Waterloo, 319-234-3075

Central C&D Recycling, Des Moines, 515-243-6402

Fuller Salvage and Wrecking, Waterloo, 391-233-2546

Home Recycling Exchange, Des Moines, 515-282-9296

House and Garden Restoration Specialties, Des Moines, 515-243-3985

Iowa Demolition and Recycling Services, Cedar Rapids, 319-393-9013
Jim's Small Demolition, Dubuque, 563-8673
Restoration Warehouse, Dubuque, 563-585-2933
Rock Creek Tree and Building Salvage, Osage, 641-732-4025
The Salvage Barn, Iowa City, ic-fhp.org, 800-541-8656

Kansas

Bahm Demolition, Silver Lake, 785-582-5190
Bob Smith Salvage, New Cambria, 785-823-8877

Kentucky

Covington Reuse Center, Covington, covingtonreusecenter.com, 859-291-0777
City Salvage and Recycling, Hopkinsville, 270-886-5606
Salvage Building Materials of Lexington, Lexington, 859-255-4700
WD Architectural Salvage, Louisville, 502-589-0670

Louisiana

Architectural Antiques Materials Company, New Orleans, 504-942-7000
Builders Antique Menagerie Co., Baton Rouge, 225-925-9582
Chandler Building Materials, Bossier City, thchandler.com, 318-747-0629
Discount Building Materials and Salvage, Covington, 985-898-2164
Gulf Coast Dismantling and Salvage, Oakdale, 318-335-9944
Krantz Recovered Wood, New Orleans, krantzrecoveredwoods.com, 504-838-6852
Louisiana Antique Woods, Lafayette, 337-269-1933
Ole Fashion Things, Lafayette, 337-234-4800
PRC/RT Salvage Store, New Orleans, prcno.org, 504-947-0038
Ricca & Puderer Demolishing & Building Materials, New Orleans, 504-656-2232
Second Chance Construction Salvage, Gretna, 504-367-7717
Sustainable Salvage, New Orleans, preno.org/shop/salvagestore.php, 504-940-4797
The Architectural Antiques Bank, New Orleans, 504-523-2702
The Wreckers Warehouse, New Orleans, 504-525-4911
White Lumber and Architectural Demolition, New Orleans, 504-486-7576
Will Branch Antique Lumber, Bogalusa, willbranch.net, 985-732-3798

Maine

Interstate Building Salvage, New Vineyard, 207-778-9340
Maine Antique Structures Salvage Company, Rockport, 207-594-0607
Maine Housing Building Materials Exchange (BME) , Lisbon, mainebme.org,
    207-636-7670

Old House Parts Co., Kennebunk, oldhouseparts.com, 207-985-1999
Portland Architectural Salvage, Portland, portlandsalvage.com, 207-780-0634
Seacoast Architectural Salvage, Rockport, 207-594-4836

Maryland

Frederick Non-Profit Building Supply Inc., Frederick, 301-662-2988
Second Chance, Baltimore, secondchanceinc.org, 410-385-1101
Tri-State Reuse Center, Hancock, 301-678-6160
Vintage Lumber Company, Woodsboro, vintagelumber.com, 301-845-2500

Massachusetts

18th & 19th Century Recycling, Spencer, oldboards.com, 508-612-0351
Atlantic Building Salvage, Ashland, 508-231-1473
Building Materials Resource Center , Roxbury, bbmc.com, 617-442-8917
Central Building Salvage Corp., Everett, 617-387-3700
JC Antique Boards and Beams, Nantucket, 508-325-8808
New England Demolition & Salvage, East Wareham, 508-291-7258
Piece by Piece Deconstruction, Amherst, 413-992-7609
Pioneer Valley Deconstruction, Springfield, 413-827-0781
Restoration Resources, Boston, restorationresources.com, 617-542-3033
RJ O'Brien Building Wrecking & Salvage, Rockland, 781-878-1961
The Olde Bostonian Architectural Antiques, Dorchester, 617-282-9300

Michigan

21st Century Salvage, Ypsilanti, 734-485-4855
Architectural Salvage Warehouse of Detroit, Detroit, aswdetroit.org, 248-245-2648
Architectural Salvage Wing-Grand Illusion, Grass Lake, 517-522-8715
Detroit Building Materials, Detroit, 313-965-6520
Heritage Architectural Salvage and Supply, Detroit, 313-345-3711
KD Used Brick & Building Material, Detroit, 313-923-4129
Larry's Building Materials, Detroit, 313-273-4699
Motorcity Building Materials Center, Detroit, 313-843-7540
Odom Re-Use Co., Grawn, odomreuse.com 231-276-6330
Recycle Ann Arbor, Ann Arbor, recycleannarbor.org, 734-662-6288

Minnesota

All State Salvage Inc., St. Paul, 651-488-6675
Bauer Brothers Salvage, Minneapolis, 612-331-9492
Better Homes and Garbage, Minneapolis, bhandgarbage.com, 612-644-9412

City Salvage Antiques, Minneapolis, 612-627-9107

Minnesota Materials Exchange, Minneapolis, mnexchange.org, 800-247-0015

Minnesota Timber Salvage, Foreston, 320-369-4507

North Shore Architectural Antiques, Two Harbors, north-shore-architectural-antiques
    .com, 218-834-0018

Old Growth Woods, St. Paul, oldgrowthwoods.com, 651-271-7544

Rural Resource Recovery, St. Paul, 651-695-1732

The Reuse Center, Minneapolis, thereusecenter.com, 612-724-2608

Mississippi

Back Road Architectural Salvage Services, Ridgeland, 601-957-3777

H&K Salvage, Gulfport, 228-832-9499

Old Mississippi Lumber, Holly Springs, 662-252-3395

T&E Salvage Buy & Sell, Biloxi, 228-392-9814

Missouri

Anderson Fine Carpentry and Salvage, Kansas City, 816-531-5976

Elmwood Reclaimed Timber, Kansas City, 816-532-0300

Missouri Deconstruction, Columbia, 573-489-4758

Planet Reuse, Kansas City (also in Chicago), planetreuse.com, 816-918-1120

ReSource St. Louis, St. Louis, resourcestlouis.org

Montana

Heritage Timber, Missoula, 406-830-3966

Home Resource, Missoula, homeresource.org, 406-541-8300

Industrial Salvage and Demolition, Missoula, 406-543-8893

Nebraska

MT Salvage, Omaha, 402-391-5315

RPM Salvage, Omaha, 402-346-4470

Nevada

Phil's Salvage Emporium, Las Vegas, 702-382-7528

New Hampshire

Admac Salvage, Littleton, admacsalvage.com, 603-444-1200

Architectural Salvage, Exeter, oldhousesalvage.com, 603-773-5635

LL&S Wood Recycling, Salem, 603-898-4098

Vermont Salvage, Manchester, 603-624-0868

New Jersey

ATS Wood Recycling, Bridgewater, 908-725-8484
Recycling the Past, Barnegat, recyclingthepast.com, 609-660-9790
Restoration Materials Company, South Plainfield, 800-336-6548
Trenton Materials Exchange, Trenton, 609-278-0033
Yanuzzi Demolition and Recycling, Orange, 973-672-8333

New Mexico

Coronado Wrecking & Salvage Co., Albuquerque, 505-877-2821

New York

A-1 Salvage, South Otselic, 315-653-4409
American Architectural Salvage and Demolition, Greenfield Center, 518-580-1849
Architectural Antiques, Rochester, 585-271-6290
Architectural Salvage Warehouse, Brooklyn, 718-388-4527
ARROW Reuse Center, Long Island, 718-472-1180
Building Materials, Catskill, 845-649-0722
Build It Green! NYC, Astoria, bignyc.org, 718-777-0132
Demolition Depot, New York, 212-860-1138
Dorp Salvage Co., Schenectady, 518-393-1744
Environmental Construction Outfitters, Bronx, 800-238-5008
Finger Lakes Reuse, Ithaca, fingerlakesreuse.org, 607-257-9699
Gothic City Architectural Antiques, Buffalo, 716-874-4479
Historic Albany Foundation Architectural Parts, Albany, 518-465-2987
Historic Home Supply, Troy, 518-266-0675
Horsefeathers Architectural Antiques, Buffalo, horsefeathers-antiques.com,
    716-882-1581
House Parts, Rochester, 716-32-2329
Hudson Valley Materials Exchange, New Windsor, hvme.com, 845-567-1445
Irreplaceable Artifacts, New York, 212-777-2900
Kleine's Antique Barnwood Flooring & Lumber, Eastport, 516-325-8955
Levanna Restoration, Auburn, 315-252-6817
Olde Good Things, New York, ogtstore.com, 212-989-8401
Rebuild WNY Inc., statewide, rebuildwny.org, 716-335-1500
ReHouse Architectural Salvage, Rochester, rehouseny.com, 585-288-3080
ReHouse Inc., Webster, 585-872-1450
Urban Archaeology, New York, 212-431-4646
Vintage Barns, Woods & Restoration, High Falls, 845-340-9870
Zaborski Emporium, Kingston, 845-338-6465

North Carolina

Antique Building Materials, High Point, 336-889-6113

Architectural Salvage, Charlotte, 704-552-7560

Architectural Salvage of Greensboro, Greensboro, 336-389-9118

Asheville Architectural Salvage, Asheville, 828-281-2600

Asheville Recyclers, Asheville, 828-254-5700

Axel Demolition & Salvage, Hillsborough, 919-644-8244

Building Supply Recycling Center, Durham, 919-490-0414

Building Supply Salvage Inc., High Point, 336-889-2207

Gideons Home Building Materials and Salvage, Greensboro, 336-294-0789

Material Salvage and Recycling, Burlington, 336-584-1193

Miller C&D Recycling, Salisbury, 704-279-2012

Orange County Landfill, Chapel Hill, 919-968-2788

Piedmont Salvage and Equipment, Greensboro, 336-510-6905

Renovator Supply of NC, Durham, 919-490-0414

Rocky's Material and Salvage, Kinston, 252-522-2424

Salvage Building Materials, Winston Salem, 336-724-1739

Salvage King, Staley

The Salvage House, Bonlee, ncagr.gov, 919-837-2376

Third Creek Salvage Company, Statesville, 704-872-7502

Vintage Beams and Timbers Architectural Salvage, Sylva, 828-586-0755

Waughton Millwork and Salvage Building Material, Winston Salem, 336-788-0990

Wilmington Architectural Salvage, Wilmington, 910-762-2511

Ohio

Architectural Artifacts, Toledo, 419-243-6916

Columbus Architectural Salvage, Columbus, 614-299-6627

Lincoln Street Salvage, Minerva, 330-868-1375

North Hill Salvage Store, Akron, 330-762-4509

Rex Salvage Store, Akron, 330-773-8605

Salvage II, Minerva, 330-868-3137

Salvage Masters, Painesville, 440-942-8769

The Stock Pile, Canton, thestockpile.org, 330-451-7446

United Salvage Co., Akron, 330-253-2403

The Valley Building Materials, Cincinnati, 513-921-2822

Oklahoma

Architectural Antiques, Oklahoma City, 405-232-0759

Dawson Building Supply, Tulsa, 918-832-0071

Oregon

1874 House Antiques, Portland, 503-233-1874
Aurora Mills Architectural Salvage, Aurora, 503-678-6083
Builders City, Portland, 503-285-0546
Cascadia Reclaimed Timber, Clackamas, cascadiareclaimedtimber.com, 971-235-0283
Craftmark Reclaimed Wood, McMinnville, craftmarkinc.com, 503-472-6929
Gorge Rebuild-it Center, Hood River, rebuildit.org, 541-387-4387
Heartwood ReSources , Roseburg, heartwoodresources.org, 541-673-4070
Knez Building Materials, Clacmus, 503-655-1991
Lovett Deconstruction, Portland, lovettdeconstruction.com, 503-286-0056
Metro Solid Waste & Recycling, Portland, 503-797-1663
Morrow's Used Building Materials, Medford, 541-770-6867
Peddle'n Pete Secondhand Store & Used Lumber, Merrill, 541-798-1037
Rejuvenation House Parts, Portland, 503-238-1900
The Rebuilding Center, Portland, rebuildingcenter.org, 503-331-1877
The Timber Recycler, Eugene, 541-687-0817

Pennsylvania

Architectural Antiques Exchange, Philadelphia, architecturalantiques.com, 215-922-3669
Architectural Emporium, Canonsburg, 724-746-4301
The Barnwood Connection, Barto, 610-845-3101
Building Materials Exchange, Philadelphia, 215-423-3613
Central Salvage, Narbeth, 610-667-1186
Conklin's Authentic Antique Barnwood, Susquehanna, 717-465-3832
Construction Junction, Pittsburgh, constructionjunction.org, 412-243-5025
Delaware Valley Recycling, Philadelphia, 215-724-2244
Material Reuse, Pittsburgh, 814-571-8659
Olde Good Things, Scranton, ogtstore.com, 570-341-7668
Pittsburgh Recycling Services, Pittsburgh, 412-420-6000
ShopDemoDepot, Greensburg, shopdemodepot.com, 724-552-0491
Recycle Shack, Royersford, 484-686-7641
Russo Demolition and Salvage, Altoona, 814-946-3215
Sahd Frank Salvage, Columbia, 717-684-8506
U.S. Recycling and Wrecking, Lancaster, 717-393-2992
Zuk Lumber & Demolition, Bethlehem, 610-868-7440

Rhode Island

Gladu Wrecking & Recycling, Woonsocket, 401-769-7792

New England Architectural Center, Warwick, newenglandarchitecturalcenter.com, 401-732-1383

South Carolina

Antique Hardware and Home, Buffington, 800-422-9982
Browns Housewrecking & Salvage, Charleston, 843-722-1643
H&H Fencing and Salvage, Columbia, 803-736-6631
Old House Salvage, Piedmont, theoldhousesalvage.com, 864-243-5990
Wild Clover Reclamation & Lumber Co. Inc., Murrells Inlet, 803-237-5490

South Dakota

Architectural Elements, Sioux Falls, 605-339-9646
Materials Clearance and Salvage, Rapid City, 605-343-1993
Second Chance Lumber, Viborg, 605-766-5145

Vermont

ReNew Building Material & Salvage, Inc., Brattleboro, renewsalvage.org, 802-246-2400

Tennessee

Architectural Exchange, Chattanooga, 423-697-1243
Bob's Salvage, Jackson, 731-668-9431
Burnett Demolition and Salvage, Knoxville, 865-637-3996
Hailey Salvage and Building Materials, Nashville, 615-224-9050
Loudon County Salvage and Building, Knoxville, 423-525-1926
Nashville Discount Building Materials, Nashville, 615-292-7856
Salvage and Building Materials, Clinton, 865-457-7897
Salvage Lumber Co., Knoxville, 865-525-6645
Southeastern Salvage, Chattanooga, 423-892-5766

Texas

AAA Salvage & Demolition, Dallas, 214-428-1888
Abe's Salvage, Fort Worth, 817-536-2381
Adkins Architectural Antiques, Houston, 713-522-6547
Architectural Antiques Salvage, San Antonio, sa-antiques.com, 210-226-6863
Building Materials Outlet, Austin, 512-836-1663
Discount Home Warehouse, Dallas, dhwsalvage.com, 214-631-2755
Discovery Architectural Antiques, Gonzales, 830-672-2428
Moore for Less Salvage Discount Building Materials, Rhome, 817-636-2552

Old Lumber Yard Antiques, Ponder, 940-479-0203
Pieces of the Past, Austin and San Antonio, pieces-of-the-past.com, 512-326-5141
Salvage Lumber of Texas, West, 254-826-4458
Salvage Sale Inc., Houston, 713-286-4601
Scott's Salvage, Waco, 254-829-1448
Second Chance Building Components, Columbus, 409-732-6646
Union Salvage, Seagoville, 936-560-4534
Welpman & Son Door & Salvage, Fort Worth, 817-535-0906

Utah

Demolition Salvage Supply Company, Salt Lake City, 801-539-1140
George's Demolition & Salvage, Salt Lake City, 801-521-8717

Vermont

Architectural Salvage Warehouse, Burlington, architecturalsalvagevt.com,
    802-658-5011
Mason Brothers Architectural Salvage, Essex Junction, 802-879-4221
ReCycle North, Burlington, 802-658-4143
ReNew Building Materials & Salvage Inc., Brattleboro, 802-246-2400
Second Harvest Antique Lumbers, Jeffersonville, 802-644-8169
Vermont Salvage Exchange, White River Junction, vermontsalvage.com,
    802-295-7616

Virginia/Washington DC

Antique Building Products, Amherst, antiquebuildingproducts.com, 804-946-0634
Black Dog Salvage, Roanoke, blackdogsalvage.com, 540-343-6200
Caravati's Architectural Antiques, Richmond, caravatis.com, 804-232-4172
Empire Salvage & Recycling Inc., Bluefield, 276-322-3554
Governor's Antiques and Architectural Supply, Mechanicsville, 804-746-1030
Green Recycling Network, Northern Virginia, DC, Maryland, greenrecyclingnetwork
    .com, 703-880-8428
Hamilton Salvage Building Materials, Coeburn, 276-762-5140
Imperial Building Supply, Norfolk, 757-489-4254
ReBuild, Springfield, rebuildwarehouse.org, 703-658-8840
Virginia Antique Building Materials, Pulaski, 540-980-4232

Washington

2Good2Toss, statewide, 2good2toss.com, 509-575-2782
Northwest Salvage & Second Hand, Vancouver, 360-694-0662

Olympia Salvage, Olympia, 360-705-1330
Rabanco Recycling, Seattle, 425-646-2476
Ray's Demolition Warehouse, Spokane, 509-533-1903
Reusable Building Materials Exchange, Seattle, 800-325-6165
Seattle Building Salvage, Seattle, 425-374-2550
Second Use, Seattle, seconduse.com, 206-763-6929
Sound Builders Resource, Olympia, 360-753-1575
Waste Not Want Not, Port Angeles and Port Townsend, 360-379-6838

West Virginia

Futineer's Wood Recycling, Lost Creek, 304-622-0535
Southland Surplus Building Materials, Beckley, 304-252-6515

Wisconsin

American Resource Recovery, Milwaukee, 414-355-8500
Beaver Wrecking and Salvage, Beaver Dam, 920-887-7030
DeConstruction Inc., Madison, 608-244-8759
Homesource Center, Milwaukee, 414-344-4142
Old House Salvage, Wausau, 715-849-5077
Reclaimed Lumber Company, Waukesha, 262-798-8986
Salvage Heaven Inc., Milwaukee, 414-329-7170
Scarboro River Barn and Lumber, Green Bay, 920-498-1755
WasteCap Wisconsin, Milwaukee, 414-961-1100

Books

Beautiful and Abundant, Bryan Welch
Building Green: A Complete How-to to Alternative Building Methods (Earth Plaster, Straw Bale, Cordwood, Cob, Living Roofs), Clarke Snell and Tim Callahan
Building a Straw Bale House: The Red Feather Construction Handbook, Nathaniel Corum
Buildings of Earth and Straw, Bruce King
Little House on a Small Planet, Shay Salomon
The Not So Big House, Sarah Susanka
Redux: Designs That Reuse, Recycle, and Reveal, Jennifer Roberts
Serious Straw Bale: A Home Construction Guide for All Climates, Paul Lacinski and Michel Bergeron
Simply Imperfect: Revisiting the Wabi-Sabi House, Robyn Griggs Lawrence
Small Strawbale: Natural Homes, Projects & Designs, Bill Steen, Athena Swentzell Steen and Wayne J. Bingham

Straw Bale Building: How to Plan, Design and Build with Straw, Chris Magwood and
Peter Mack
Straw Bale Construction Details: A Sourcebook, Ken Haggard and Scott Clark
The Straw Bale House (A Real Goods Independent Living Book), Athena Swentzell Steen,
Bill Steen and David Bainbridge
Unbuilding: Salvaging the Architectural Treasures of Unwanted Houses, Bob Falk and
Brad Guy

Building Instructions/Information

Build It Green, Oakland, California, builditgreen.org, 510-590-3360, promotes healthy,
energy- and resource-efficient homes in California
California Straw Bale Association, Angels Camp, strawbuilding.org, 209-785-7077
Colorado Straw Bale Association, Gunnison, coloradostrawbale.org, 970-497-9575
Design-Build-Live, Austin, Texas, designbuildlive.org, 512-478-9033
Development Center for Appropriate Technology, Tucson, Arizona, dcat.net,
520-624-6628, works to promote sustainable construction and development
DIY-Home-Building, online/Madison, Wisconsin, diy-home-building.com,
608-669-8226
Dream Green Homes, Crestone, Colorado, dreamgreenhomes.com, alternative home
plans using a variety of materials/styles
Earthship Biotecture, Taos, New Mexico, earthship.org, 575-613-4409, earthship, radi-
cally sustainable green building made with recycled materials-education, design
information and services, construction, rental, etc.
Earthship Landing, Durango, Colorado, earthships.com, personal story/blog/website
of a DIY earthship builder
Ecological Building Network, San Rafael, California, ecobuildnetwork.org,
415-987-7271, promotes education in and proliferation of green building
Our Earthship, El Paso, Texas, earthpower1.com, story of one couple who built their
own earthship
Pakistan Straw Bale and Appropriate Building, International/Truckee, California,
paksbab.org, 530-902-5516, works to provide straw bale building training to those in
need worldwide
The Straw Bale Association of Nebraska/The Last Straw Journal, Lincoln, thelaststraw
.org, 402-483-5135

Building and Deconstruction Training
Nationwide

Reuse Consulting, reuseconsulting.com, 360-201-6977
The Reuse People Deconstruction Training, deconstructiontraining.org

California

CalRecycle, statewide, calrecycle.ca.gov

Washington

Workforce Development Council, Seattle-King County, 206-448.0474, deconstruction training, seakingwdc.org

Index

About the Author

JESSICA KELLNER is Editor of *Natural Home & Garden* magazine (naturalhomeandgarden.com) and a passionate advocate of using architectural salvage to create aesthetically beautiful, low-cost housing. She writes and speaks about practical, easy and inexpensive solutions for healthier and more environmentally sound homes and lifestyles, with an emphasis on the values of recycling and using reclaimed materials. Inspired by common sense ways to achieve the good life, Jessica loves living and working in the overlap between modern technology and ancient wisdom.

If you have enjoyed *Housing Reclaimed*, you might also enjoy other

BOOKS TO BUILD A NEW SOCIETY

Our books provide positive solutions for people who want to make a difference. We specialize in:

**Sustainable Living • Green Building • Peak Oil
Renewable Energy • Environment & Economy
Natural Building & Appropriate Technology
Progressive Leadership • Resistance and Community
Educational & Parenting Resources**

New Society Publishers

ENVIRONMENTAL BENEFITS STATEMENT

New Society Publishers has chosen to produce this book on recycled paper made with **100% post consumer waste**, processed chlorine free, and old growth free.

For every 5,000 books printed, New Society saves the following resources:[1]

28	Trees
2,581	Pounds of Solid Waste
2,839	Gallons of Water
3,703	Kilowatt Hours of Electricity
4,691	Pounds of Greenhouse Gases
20	Pounds of HAPs, VOCs, and AOX Combined
7	Cubic Yards of Landfill Space

[1]Environmental benefits are calculated based on research done by the Environmental Defense Fund and other members of the Paper Task Force who study the environmental impacts of the paper industry.

For a full list of NSP's titles, please call 1-800-567-6772 *or check out our website* at:

www.newsociety.com

NEW SOCIETY PUBLISHERS